WITHDRAWN FROM STOCK
DUBLIN CITY PUBLIC LIBRARIES

Leabharlann na Cabraí
Cabra Library
01-2228317

Female Innovators Who Changed Our World

Female Innovators Who Changed Our World

How Women Shaped STEM

Emma Shimizu

First published in Great Britain in 2022 by
Pen & Sword History
An imprint of
Pen & Sword Books Ltd
Yorkshire – Philadelphia

Copyright © Emma Shimizu 2022

ISBN 978 1 52678 969 3

The right of Emma Shimizu to be identified as Author of this work has been asserted by her in accordance with the Copyright, Designs and Patents Act 1988.

A CIP catalogue record for this book is
available from the British Library.

All rights reserved. No part of this book may be reproduced or transmitted in any form or by any means, electronic or mechanical including photocopying, recording or by any information storage and retrieval system, without permission from the Publisher in writing.

Printed and bound in the UK by CPI Group (UK) Ltd,
Croydon, CR0 4YY.

Pen & Sword Books Limited incorporates the imprints of Atlas, Archaeology, Aviation, Discovery, Family History, Fiction, History, Maritime, Military, Military Classics, Politics, Select, Transport, True Crime, Air World, Frontline Publishing, Leo Cooper, Remember When, Seaforth Publishing, The Praetorian Press, Wharncliffe Local History, Wharncliffe Transport, Wharncliffe True Crime and White Owl.

For a complete list of Pen & Sword titles please contact

PEN & SWORD BOOKS LIMITED
47 Church Street, Barnsley, South Yorkshire, S70 2AS, England
E-mail: enquiries@pen-and-sword.co.uk
Website: www.pen-and-sword.co.uk

Or

PEN AND SWORD BOOKS
1950 Lawrence Rd, Havertown, PA 19083, USA
E-mail: Uspen-and-sword@casematepublishers.com
Website: www.penandswordbooks.com

Contents

Acknowledgements vii
Preface viii

Chapter 1 World Defining Moments in History 1
Joan Clarke (1917–1996) 1
Rosalind Franklin (1920–1958) 3
Ada Lovelace (1815–1852) 6
Lise Meitner (1878–1968) 10
Rufaida Al-Aslamia (Born 620) 15

Chapter 2 Our Day-to-Day Lives 17
Alice Parker (1885–?) 19
Josephine Cochrane (1839–1913) 20
Mary Anderson (1866–1953) 21
Mary Beatrice Davidson Kenner (1912–2006) 22
Stephanie Kwolek (1923–2004) 24
Ruby Hirose (1904–1960) 25
Olive Dennis (1885–1957) 28
Sutayta Al-Mahāmali (920–987) 29

Chapter 3 Improving Lives, Equality and Justice 31
Margaret Sanger (1879–1966) 31
Anne McLaren (1927–2007) 36
Laura Bassi (1711–1778) 39
Alyce Gullattee (1928–2020) 41

Chapter 4 Global Health 44
Patricia Bath (1942–2019) 45
Kin Yamei (1864–1934) 49
Kamala Sohonie (1912–1998) 54

Jane Cooke Wright (1919–2013)   58
Alice Ball (1892–1916)   60
Tu Youyou (1930–)   63
Françoise Barré-Sinoussi (1947–)   66
Rosalyn Yalow (1921–2011)   69

Chapter 5   Protecting the Earth   73
Saruhashi Katsuko (1920–2007)   73
Rachel Carson (1907–1964)   75
Wangari Maathai (1940–2011)   80

Chapter 6   Influential Projects and Leadership in STEM   84
Kate Gleason (1885–1933)   84
Caroline Haslett (1865–1957)   90
Lillian Gilbreth (1878–1972)   94
Emily Roebling (1843–1903)   100

Chapter 7   Healthcare for Children   103
Anna Freud (1895–1982)   103
Mamie Phipps Clark (1917–1983)   106
Virginia Apgar (1909–1974)   108
Helen Taussig (1898–1986)   109

Chapter 8   Understanding Our World   112
Inge Lehmann (1888–1993)   112
Maria Goeppert Mayer (1906–1972)   115
Marie and Irène Curie (1867–1934) and (1897–1956)   120
Gerty Cori (1896–1957)   123
Annie Easley (1933–2011)   125
Constance Tipper (1894–1995)   128
Marie Gayler (1891–1976)   128
Marion McQuillan (1921–1998)   128

Notes   133
Index   154

Acknowledgements

Trying to select such a small number of women to write about and do them justice has been an immensely difficult task, but writing about their lives has been an absolute privilege. Thank you to Aileen for believing in me, and to Lauren without whom we would not have met.

I have been humbled by my friends' willingness to offer their time and share their unique and important perspectives. As the primary aim of this book is to inspire, I have taken every effort to ask for advice in any areas that need to be inclusive and representative. I could not have achieved this without these friends: Matt, Mita, George, Luca, Pengk, Alice, Abi, thank you. Steve and Lauren, I salute your patience. There is no way to express my gratitude for the hours you have put in to this book.

Thank you to my families, the Greens, Shimizus, and Reisses, for always being encouraging and taking an interest. Special thanks to Baba, Sally, and my parents, Andrew and Naomi; you have supported me in my endeavours ever since I can remember.

And to my husband, Rob. I've spent hundreds of hours researching the lives, careers and romances of women in science, technology, engineering, and mathematics (STEM). Never in my wildest dreams could I have wished for anyone better than you.

Preface

When I set out to write this book at the beginning of 2020, there was no way of knowing the extent of the disruption and suffering the Coronavirus pandemic would have on the world. Having a scientific background, I have always had an appreciation for the way in which STEM shapes our lives, but it wasn't until virologists seemed to have a daily slot on the radio that it truly hit home. Public interest in science and technology inevitably increases when it drives the response to health crises, but COVID-19 has had an unprecedented social, economic and political effect that will remain in the memories of people who lived through the pandemic. At the time of writing, Dr June Raine is the CEO of the Medicines and Healthcare products Regulatory Agency (MHRA) of the United Kingdom and as such she has had a high level of responsibility in preparing for the UK's vaccination response.

Upon embarking on my research, I was delighted to find that history has no shortage of successful and respected women like Raine. Whilst the scope of this book exclusively focuses on the history of women in STEM, it is important to clarify that their work was not merely celebrated because of their gender but because it stood up to the scrutiny of the international scientific community. There are so many women who fit this description that it was a real challenge deciding on who to include. Ultimately, holding on to the belief that the most powerful inspiration comes from people and situations we can relate to, the subjects were chosen because their work had concrete relevance to our twenty-first-century lives. The incredible achievements and journeys of the forty-six women in this book really spoke to me as a woman who studied science, trained as an engineer and works in manufacturing. Although many of them may have carried out their work hundreds of miles away from where you are sitting, several lifetimes ago, they were human and faced struggles and challenges just like we do.

As each generation continues to develop and define the action required to create a more inclusive world, it is more important than ever to look back at the resilient innovators who came before us. Many of us are privileged to live in a world that has come a long way since these women were at the peak of their careers. When it comes to providing equal opportunity regardless of gender, race, sexuality or economic status, whilst promoting ethical and environmentally sustainable practices, both the global north and the global south have a long way to go.

Science, technology, engineering and mathematics have always played a major role in shaping civilisations, but there is inevitably a lag time before new advancements benefit the majority. Technological progress will usually be limited and have unanticipated consequences if social development is not considered alongside it. This is one of the reasons it is so vital for women from a diverse range of backgrounds to pursue careers that have the ability to change our lives. Otherwise, who exactly are things being designed to benefit? Caroline Criado Perez highlights one such example in her book, *Invisible Women: Exposing Data Bias in a World Designed for Men*, when she points out that when cars are crash-tested, only the male anatomy is considered.[1] It might not seem like a big problem, until she then reports that the fatality risk for women is 17 per cent higher than men of the same age placed under similar conditions.[2]

The women in this book had a hand in shaping the world, and our lives are better for it.

Whether you work in a STEM field, aspire to do so, have a family member or friend who wants to, or wish to know more about the true history of science and technology, I hope you enjoy learning about these fascinating trailblazers. Some have been denied education, whilst others were thrown out of their countries, orphaned, suffered illnesses, or were spurned because of their ethnicity or culture. And yet, they have gone down in history. Their stories are there waiting to be told and to spark discussion. To instil confidence. To inspire.

Copyright Disclaimer

The author of this book has made every effort to find and contact the original copyright holder of images presented. The author has only used images for which agreement has been given by the copyright holder or where it has been confirmed that the image is in public domain. If you believe that you are the copyright holder of any of the images presented, please contact the publisher for more details.

Chapter 1

World Defining Moments in History

Joan Clarke (1917–1996)
Mathematics, Cryptanalysis
England

If Joan Clarke had not worked alongside Alan Turing as part of the team who cracked the Enigma code at Bletchley Park, our world would be a very different place. Whilst it would be an oversimplification to say that code breakers led the allied forces to victory, Sir Harry Hinsley, an expert on British intelligence did say that the work done at Bletchley Park was responsible for ending disruption caused by German U-boats and therefore instrumental in restoring food security for citizens of the UK.[1]

Clarke was born on 24 June 1917 and attended Dulwich High School for Girls. She won a scholarship to Newnham College, Cambridge[2] to study mathematics and excelled to the point where she became the highest-scoring student on her course, earning the title of 'Wrangler'. Because the University of Cambridge did not award degrees to women at the time,[3] Clarke received no official qualification. However, her talent did not go unnoticed. As the Second World War broke out, it became obvious to those in the know that German coding methods had become more sophisticated since the First World War and that the British would need to pull together as many intelligent minds as possible to combat this. The severity of the challenge left little room for gender preference and as it became acceptable for women to be hired for classified government projects, Clarke was interviewed for a position at the UK Government Code and Cypher School (GC&CS) by her former geometry supervisor, Gordon Welchman. Details of the advertised role were deliberately vague, but Clarke took a leap of faith and accepted the job. All she knew was that mathematicians seemed to excel at the place she was going to.[4]

Once settled into to her work environment in Hut 8, Clarke was introduced to a bombe – a machine that aided the deciphering of messages encrypted by the Nazis' Enigma machine. Turing had developed this for GC&CS the year before from an earlier Polish design, and Clarke's initial task was to use it on messages that had been intercepted. This was a complex and repetitive task as the Enigma code was changed every twenty-four hours.

The bombe machine, along with the rest of the work done at Bletchley Park was praised by Prime Minister Winston Churchill for contributing to the shortening of the war and saving countless lives. Developed from a Polish device of a similar name, it worked on a number of features that were known about the German Enigma machine. The most critical of these was that an Enigma machine would never encode a letter as itself, so an 'A' in the message would never be an 'A' in the cypher. Additionally, connections made on the plugboard meant that pairs of letters were linked; if the 'W' and 'T' were connected then 'T' could not be connected to anything else. The bombe used these flaws to break the code. It would rattle through different combinations of starting position and plugboard connections based on a 'menu' of rotor orders, looking for contradictions. If it found a contradiction then that combination was known to be wrong; if no contradiction was found then it was understood that those settings were a possibility.[5] Clarke became so knowledgeable about the machine and the process that she developed ways to increase efficiency, speeding up the process of cracking codes and deciphering messages.[6]

Clarke quickly gained responsibilities; although three-quarters of employees at Bletchley Park were female,[7] she was unusual in that she worked in an office with men and early on in her employment at Hut 8 was trusted to work the night shift alone. Her senior position did not cause any day-to-day challenges but seemed to cause issues with payroll, primarily because the pay grade for 'female cryptanalyst' did not exist. Instead, Clarke was listed under the 'linguist' grade, despite none of her work involving translation. Even when she was promoted to deputy head of her hut, Clarke still received a lower salary than the rest of her male colleagues.[8]

The wartime 'goings on' at Bletchley Park have now been declassified,[9] which is why we are able to learn about and celebrate the significant contributions made by Joan Clarke. After the end of the Second World War, many women who had served at Bletchley Park returned to their previous, more domestic, lifestyles. However, Clarke's experiences in Hut 8 had introduced her to the type of job prospects open to women who excelled at logical thinking.

She was briefly engaged to Turing during her time at Bletchley Park, however it was whilst at her post-war job at GCHQ, the successor to GC&CS and Bletchley Park, that she met her future husband, Lieutenant Colonel John Kenneth Ronald Murray. Apart from taking a break due to Murray's ill health, Clarke worked until her retirement in 1977.

Clarke passed away aged 79 in her house in Headington, where an Oxfordshire Blue Plaque has been displayed since 2019 as a reminder of this talented woman who so enthusiastically gave her brilliant mind to help her country.

Rosalind Franklin (1920–1958)
Chemistry
England

The story of Rosalind Franklin is a tough one to stomach. Ever since her under-publicised contribution to the discovery of DNA that earned Crick, Watson, and Wilkins a Nobel Prize came to light, it has been all too easy to focus on the negativity and unfairness that clouded her short life. It is often easy to simplify and vilify, but in reality, most scientific discoveries and advancements come about because of a team effort. It can be difficult to fairly apportion credit surrounding innovation; there may be the person who makes the initial discovery, then the person who finds an explanation for it, followed by the person who develops the theory or finds a useful application. Unfortunately, it is impossible to change history but whatever controversies surround the complex detail of her time as a researcher at King's College London we should never forget the absolute gem of a gift that Rosalind Franklin gave humanity: the work that, once shared, started the domino effect that led to our current understanding of the structure and role of DNA and, perhaps even more importantly, the role it plays in diseases and medical conditions.

Born in London in 1920, Franklin was determined and headstrong, both useful qualities for women who strived for equal opportunities and wished to have their work taken as seriously as their male peers. She excelled at school and entered the same Cambridge College as Joan Clarke (p.1). But whilst Clarke's interests lay in mathematics, Franklin was interested in physical chemistry. By the time she had finished her undergraduate degree course, Britain was in the middle of the Second World War. This changed university life considerably, with staff being either called away to contribute to the war effort or detained if they were German. The arrival of refugee and French scientist Adrienne Weill at Cambridge was a significant event in Franklin's life[1] as she gained a mentor and friend who would support her for years to come. To aid her country during the rest of the war, Franklin investigated the structure of coal, specifically the effect of temperature and carbon content on the size of pores.[2] The larger the pores, the larger the molecules that can pass through the material, in the same way that small seeds can pass through a colander but not a sieve. By providing an understanding of its microstructure, Franklin's work helped in classifying types of coal and predicting their efficiency when used as fuel. Another important property of coal is its ability to react with and adsorb liquid or gas molecules, which is useful when considering applications such as gas masks. Although this research was conducted whilst working for the British Coal Utilisation Research Association, she submitted her findings to the University of Cambridge and received a PhD.

With the help of Weill, Franklin embarked on an exciting new chapter of her life after the war, moving to Paris in 1947 to work at the Laboratoire Central des Services Chimiques de l'État. Her laboratory leader, Jacques Mering, was skilled at an experimental technique that used an X-ray split into different directions; by measuring the angles and intensity of these beams, the atomic and molecular structure could be observed. Franklin applied this to her ongoing studies of coal and materials that contained carbon and she became an expert at X-ray crystallography.

Having received a fellowship, Franklin returned to England in 1951 and started working at King's College London. She continued to use her talents as a crystallographer but at the request of the director of the Medical Research Council's Biophysics Unit, swapped her work on coal for DNA research. Unfortunately, Franklin's new position got off to a bad start even before her first day in the laboratory. Unbeknownst to her, there was already a physicist researching DNA at the same institution; Maurice Wilkins and his PhD student Raymond Gosling had advanced the study of the elusive molecule to the point where they had managed to take some photographs of it.[3] Franklin was of the opinion that she would be the sole scientist working on the X-ray diffraction of DNA. Wilkins on the other hand, was not aware that Franklin had been appointed to take over this part of the research and that Gosling was to become her assistant. Thus, their relationship had a rocky beginning and became worse over time when Wilkins realised that his new colleague would be there in more than an 'advice-giving' capacity.[4]

The pressure started to mount early in 1953, as it became apparent that the American chemist Linus Pauling had a keen interest in DNA.[5] Whilst the group at King's ramped up their efforts to obtain significant X-ray diffraction data, James Watson and Francis Crick were working on a physical model at Cambridge University. The problem faced by Watson and Crick was trying to line up the principles that the bases involved in the construction of a DNA molecule were effectively random, but somehow the X-ray diffraction data showed that molecule could crystallise.[6] This indicated that some part of the structure exhibited regularity. Franklin had recorded a lot of relevant data and had included it in an informal report which she showed to Max Perutz, head of the Molecular Biology Unit at the Cavendish Laboratory. He in turn passed this to Sir Lawrence Bragg, who showed it to his colleagues Watson and Crick. None of this information was confidential as it had been previously presented by Franklin herself at a seminar held at King's [7]. Watson had been present at the time, though had not understood the implications.

Meanwhile, Gosling had taken an image called 'Photo 51'[8] which was to go down in history. In a turn of events that has caused much controversy over the years, Wilkins decided to show this image to Watson when the Cambridge researcher was visiting King's.[9]

Franklin was gradually realising that DNA had a double helical structure with complementary bases. However, she was not able to provide a satisfactory mathematical model to back this up. Franklin's data, combined with 'Photo 51', had made a significant contribution to Watson and Crick's understanding of the molecule and by March 1953 they had created a physical model. Upon its completion, Franklin and Wilkins were invited to Cambridge to review the breakthrough; both agreed that the model was correct.[10] Given that this had taken much effort from both groups at Cambridge and King's, it was decided that Watson and Crick were to publish their paper on the structure of DNA.[11] This was to be followed up by Franklin and Wilkins' separate publications of the supporting data. All three papers were published in quick succession in the journal *Nature* in April 1953. However, by the time the papers were published, Franklin's discontent with the working environment had caused her to leave King's and her work on DNA.

When Franklin died of ovarian cancer at the age of 37, she was known mainly for her work on viruses. She had been able to isolate the infectious part of the virus that caused Tobacco Mosaic disease.[12] The timing of her early death in 1958 possibly led to the perceived academic snub that Franklin's legacy sadly became defined by. The delay between the discovery being published and the Nobel Prize being awarded in 1962 meant that Franklin fell foul of the rule that the prize cannot be awarded posthumously.

In conclusion, there are several aspects of her time at King's that culminated in the events of the often-sensationalised story of Franklin and the double helix. Her strong character and rocky professional relationships led to her poor treatment and undoubtedly made her work environment at King's very unpleasant. Secondly, there was clearly some mismanagement and poor communication surrounding the research interests of Franklin and Wilkins that led to ill feeling, even before they started collaborating. Additionally, lack of formal procedure regarding the sharing of information between universities led to the incident of 'Photo 51' being messier than it needed to be.

Most important was the commitment to scientific advancement and quality of research. Franklin achieved incredible results, despite her solitude and lack of strong team connection. Because of her exacting standards she also chose to defer from publishing some aspects of her interpretation, rather than putting out a theory that could not be rigorously backed up with maths and chemistry. This does her credit as a scientist; truth should be held in higher esteem than any promise of public recognition or professional prestige.

Ada Lovelace (1815–1852)
Mathematics
England

Ada Lovelace's path into science was an unusual one. Her father was the famous poet, Lord Bryon, who was notoriously troubled and an unashamed womaniser. He and his wife separated when Lovelace was just five weeks old. Lady Byron remained bitter about her husband's character for a long time after their separation and insisted that her daughter be subjected to a rigorous mathematical education in the hope that developing a logical mind would ward off any 'madness' she might inherit from her father. Despite her good education, Lovelace's childhood was full of challenges; her health was not good and she was constantly watched over by her mother's friends, who were looking for signs of immorality. Despite this façade of motherly concern, Lady Byron's true feelings for her daughter may be more accurately portrayed in a letter in which she referred to Lovelace as 'it'.[1] Fortunately, the girl was able to form positive relationships with her maternal grandmother and with her tutor, Mary Somerville.

By the age of 17, Lovelace was intelligent enough to understand one of the latest technological developments, the difference engine, invented by Charles Babbage. She was introduced to the inventor in the summer of 1833 and took great interest in this machine, which when recreated by the Science Museum in 1991 to commemorate the bicentenary of Babbage's birth, measured 3.4m long and 2.1m high.[2] The concept of using mechanical parts to complete calculations must have been the equivalent of witnessing the latest in artificial intelligence; Lovelace noted with excitement how she watched the 'thinking machine' calculate the second and third powers (x^2 and x^3) of several numbers, as well as find the root of a quadratic equation.[3] However, what Lovelace saw at this time was in fact only one-seventh of the machine Babbage had designed. Despite a fair amount of government investment it wasn't long before he appeared to give up and move on to his next project, which Lovelace would come to know very intimately. Whilst the difference engine had aimed to produce tables of calculations that were time-consuming to do by hand, his new analytical engine was not so niche, and its similarity to the technology we use today is the reason Babbage is often called 'the father of the computer'.

In 1840, Babbage gave a lecture in Turin, Italy. Perhaps sensing that the inventor was on the cusp of something world-changing, the engineer (and future prime minister of Italy) Luigi Menabrea translated the lecture notes into French. Lovelace was then tasked with translating this version into English, which she did with great competence. After completing the translation, she also added her own comprehensive notes, with points listed in alphabetical order, A to G.

There is some speculation as to whether Lovelace herself played a part in programming the analytical machine, which employed the same principle as the loom cards used to make intricate patterns in lace. This controversy aside, perhaps one of the reasons Lovelace made such a mark on society of the time, the development of computers that followed, and science in general, was her unique ability to think creatively and not be restricted by the details of the science that she loved so much. Whilst her male peers were obsessing over the niche applications for calculation that Babbage's machine offered, Lovelace (correctly) predicted that it could be used for a wider variety of things that would greatly improve the world. Rather than seeing a glorified calculator, Lovelace saw a machine that could be used for various applications, including the composition of music. Doron Swade, an expert in the history of computing and the man responsible for building the physical manifestation of Babbage's unfinished difference engine, described her as 'a prophet of the computing age'.[4] Perhaps she was able to think laterally because she had the luxury of not being motivated by profit. Perhaps she inherited a curious, poetic mind from her father, whom her mother was so keen to erase from her very being. Those of us who have the inclination to go into science are not always destined to make major discoveries, invent things, or become pioneers in our fields. But, like Lovelace, we should show passion for what we do, communicate well and make steps, no matter how big or small, to inspire future generations.

We may never fully know the extent of her contribution, but there is no denying that Lovelace was incredibly intelligent and an absolute visionary amongst her female peers. She didn't mind being seen as 'odd' for discussing equations with men, at a time when the rest of aristocratic society was preoccupied with gossip and focusing on trying to keep up with the latest fashion. Whatever her contribution, throughout history she has instilled in girls and women the belief that they are capable of understanding and working in mathematics, engineering, and computer science. Her legacy is so great that the US Department of Defence commissioned a computer language to be named 'Ada' in her honour, and many awards and medals also use her name to recognise achievements. Even colleges and buildings have been named after her.

A technically minded female being so celebrated can only be a good step in the direction of improving STEM opportunities for girls and women. Lovelace was followed by many inspirational women who shared her appreciation for how technology could shape the world.

Hollywood actress Hedy Lamarr had an innovative mind and during the Second World War set about trying to solve the issue of radio-controlled torpedoes getting jammed and set off course. With musician and composer George Antheil, she co-invented a device that incorporated a frequency-hopping signal. Whilst it was not used by the US navy at the time, the frequency-hopping spread spectrum that their device incorporated proved useful for military purposes in the 1960s, just years after the original patent had expired. A more modern application is low-energy protocols such as Bluetooth.

Grace Hopper was another American pioneer who laid the foundations of technology that shaped the twenty-first century. Hopper was one of the first programmers of the Harvard Mark I computer and also one of the first people to create a linker. Her version was a computer system program that converted English into code. Like Lovelace, Hopper struggled to have her suggestions taken seriously because the general opinion was that computers were only useful for conducting calculations. After persisting for several years, she persuaded her colleagues to see the merits of writing programs in English and a computer language for data processors called common business-oriented language (COBOL) was eventually developed. Decades later, computer scientist Karen Spärck Jones would make further advancements to the way in which we interact with computers. Spärck Jones developed the concept of term frequency-inverse document frequency (tf-idf), which is a statistic that determines how important a word is to a single document that is part of a larger set of texts. This development means that we can now use search engines safe in the knowledge that we will only be shown results that are relevant to our search term.

Many cultures have moved on from the days in which women were banned from laboratories or had any kind of autonomy over their scientific work. However, with the digital revolution showing no sign of slowing, humanity is in grave danger of repeating past mistakes. The technology gap is yet another aspect of life that is starting to divide genders, economic classes, generations and nationalities due to lack of opportunity and diversity in some areas. When a new technology is designed and created by a small subset of its intended users, there are often issues with accessibility. For example, facial recognition algorithms are not universally accurate; studies have shown that accuracy is lower for populations that don't coincide with the images that the algorithm is trained on.[5] Unsurprisingly, this can be problematic when used in the criminal justice system as the algorithms are primarily tested on older, Caucasian faces, leading to difficulties telling younger black people apart.

In terms of technology, our world is vastly different to the one Ada Lovelace knew. Given the advancements made during the last 200 years, it is humbling to think of what will be available to humankind given another two centuries of scientific discovery and research. Current and future generations have the opportunity and responsibility to develop technology in a conscious way. It is important that new designs take into account the people and habitats affected by the collection of resources required and, like Lovelace, those with influence should drive innovation in a way that doesn't exclude those not considered to be the primary user. Sustainable technology for the many and not the few has the power to change the world.

Lise Meitner (1878–1968)
Physics
Austria-Hungary, Germany, Sweden

Planet Earth is facing an energy crisis. As the global population grows and dependence on ever-increasing energy consumption to support day-to-day living becomes inevitable for more people, there is a critical need to continue the development of technology that allows us to move away from burning fossil fuels. One particular woman in STEM paved the way in this regard, by describing how a heavy atomic nucleus can be broken down into lighter nuclei, releasing energy in the process. Put simply, Lise Meitner contributed to the understanding of nuclear fission, a reaction that can be harnessed to generate energy in a cleaner way than burning fuel.

Born on 7 November 1878,[1] Meitner had an intellectual upbringing. Although of Jewish origin, her family converted to Protestantism; whilst religion was important, logic, music, and critical thinking were disciplines also held in high regard. Meitner's desires were counter-cultural to say the least. Shy but determined, she wished to receive a broader and deeper education than the bare minimum that the girls of Viennese society had access to. When a female student reached the age of 14, her education was deemed to be complete.[2] Far from the operas, balls and suitors that captured the interest of her peers, Meitner's dreams involved studying science at the University of Vienna.

As girls were not able to attend high schools that offered university preparation, she turned to her father for help. Both parents wanted their daughter to have an insurance option, so agreed that she could attempt the university entrance exams on the condition that she also obtain a teaching qualification that would offer job security as a back-up. The exams were so difficult that the usual study period was eight years.[3] Meitner managed it in two, and was one of only four female students who achieved good enough grades to secure their admission to the University of Vienna. Shifting into an academic environment where learning was the meaning of life, Meitner thrived. She now had role models and tutors who could further inspire her interest in physics. Her thesis was published in 1906, after being reviewed by academics who included the renowned physicist Ludwig Boltzmann, making her one of the first women to earn a PhD from the University of Vienna. Her association with scientists whose names pepper textbooks to this day was just beginning. Having proved her ability to conduct independent research, Meitner moved to Berlin's Friedrich Wilhelm University. Max Planck invited her to attend lectures in his own home and, hoping to get involved with some research, she met with Heinrich Rubens, the head of the experimental physics institute. It was suggested that she join Otto Hahn at the Chemistry Institute; when Meitner and Hahn became acquainted shortly after this meeting, neither one could have anticipated the decades-long partnership that was about to unfold.

Radioactivity was a buzzword of the day and Hahn and Meitner worked successfully by combining his knowledge of chemistry and her expertise in physics to add to the bank of information that was steadily growing around the topic. They started studying beta radiation, discovering new radioactive substances in the process. As the concept of professional female scientists was extremely new and not entirely welcomed by everyone at the institute, Meitner faced several barriers including lack of toilets, and was restricted access to the rest of the building. When positions became available at the newly established Kaiser Wilhelm Institute (KWI) for Chemistry in 1912, both Meitner and Hahn moved. Their work continued to be fruitful; they published steadily, with a high point being their discovery of the element protactinium in 1918. A few years after this, their research interests went separate ways. Hahn was first and foremost a chemist, whilst Meitner was keen to delve further into the world of nuclear physics. However, their professional separation was not to last.

The rise of Hitler and the Nazi Party caused concern, but Meitner did not feel overly threatened as she had two safety nets. The KWI was not affiliated to the government so politics had not historically interfered with the science conducted there. Additionally, Meitner was Austrian which gave her immunity from being dismissed due to her status as non-Aryan. As society regressed under the oppressive regime of the Nazi Party, Meitner was at the cutting edge of science. The neutron had been discovered in 1932[4] and Meitner used this novel particle in her experiments involving nuclear reactions. Enrico Fermi had been using neutrons to bombard the known elements of the periodic table to see what artificial radioactive substances could be found.[5] This pioneering work was one of the reasons Meitner decided to approach Hahn after their years apart and suggest a collaborative project. Her primary interest was to see how the nuclear reactions could lead them to the transuranium elements, which are elements with an atomic number greater than 92 (uranium). Although Hahn took some persuading to join her on the project, Meitner got her wish, bringing chemist Fritz Strassmann along for good measure.

By 1938, the socio-political situation in Germany had become too much of a threat for Meitner to stay at the Chemistry Institute. Hahn was staunchly anti-Nazi but was put under increasing pressure to dismiss Meitner because of her Jewish heritage. She had previously requested permission to leave the country; the fact that this was rejected because the Nazis did not want Jewish people to travel abroad to represent German science was a major red flag. In a pivotal strategic move that became known as *Anschluss*, Germany annexed Austria in the spring of 1938. Because of this, Meitner lost what little protection she had; she needed to escape, and quickly. The severity of the situation was perhaps highlighted by the invitations she started receiving from colleagues abroad, urging her to make a decision sooner rather than later.[6] Given that news sources were unreliable, moving abroad without first securing a job was an unwelcome prospect for Meitner, but fear of uncertainty was quickly overshadowed when the Ministry of Education became aware of who she was. All hopes of emigrating were dashed when it transpired that her Austrian passport was invalid due to the *Anschluss*. Once physicist Niels Bohr saw the trouble she was in, a web of discrete communication began to spread across Europe as her international colleagues worked together to find her a way out of Germany, as well as a new position at Manne Siegbahn's laboratory in Sweden.

And thus began the mother of long-distance relationships. The communication was frequent and highly productive for a time; Meitner and Hahn were in contact every few days with him reporting what was happening in the laboratory, and his ex-laboratory partner weighing in with her interpretation.[7] The volatile situation in Germany meant that joint publishing was out of the question, but there is plenty of evidence in letters written from Hahn to Meitner in the run up to the discovery of fission, that any positive results that came from the experiments in Berlin would be treated as an accomplishment for Meitner, as well as himself and Strassmann.[8] Frustration was rife at this time. Meitner, now working in a laboratory in Stockholm where she had received a minimal welcome, was mourning the life and career she had been forced to leave behind in Germany. In truth, things were not going so well back at the KWI either. Hahn and Strassman had lost the physicist of their group and progress had slowed considerably. They were missing the knowledge required to explain what was causing their strange experimental results. The goal was to discover elements that had an atomic weight greater than 92, but instead, they found that the radium they were using was behaving like barium, which in fact had a lower atomic weight and number. Hahn knew that he could not hide this result that seemed to defy the laws of physics; neither could he publish without having a considered explanation for the phenomenon. His frantic letters to Meitner show how he practically begged her to help them find a way out of the professional limbo he and Strassman found themselves in. Meitner replied to Hahn's desperate letter with a sense of calm; whilst chemistry was an established science, nuclear physics was relatively new and Meitner was wise enough not to label unusual results as 'impossible'. She mused that whilst the idea of the nucleus splitting was difficult to comprehend, it could not be ruled out completely. She went on to develop this idea whilst on a walk with her nephew, also a physicist. They discussed how the nucleus could be analogous to a droplet of water; although held together by surface tension, any object that collided into it with enough force would cause the droplet to break into smaller parts. Despite the magnitude of this realisation, Meitner had mixed feelings. She had spent four years attempting to find the tranuranium elements. If her interpretation of Hahn's latest experimental results was correct, she knew that the work carried out during her last few years at the KWI had been wrong. Despite her isolation, Meitner was the key player in the Berlin team – she had initiated the project, which Hahn and Strassman joined at a later date at her request. Even after being forcibly displaced from their laboratory, Meitner continued to play a crucial role in interpreting the experimental data that her former laboratory partners were struggling to understand. With one sentence, she was able to guide the direction of the research carried out by Hahn and Strassman back in Berlin, such was their respect for her as their intellectual equal, if not leader.[9] However, due to the political situation, Hahn started to feel the consequences of his close working relationship with Meitner and withdrew from their friendship for a period. She was nominated over twelve times for the Nobel Prize,[10] but in the end, it was Otto Hahn alone who received the award. Fortunately, the full extent of Meitner's contribution came to light in the decades that followed. She received awards and honours from organisations and institutes across Europe and the US and became the first individual woman to have an element named after her. Element 109 is meitnerium.

Whilst many scientists hope that their discoveries will someday help with the advancement of humankind, unfortunately this is not always the outcome. War and destruction came to pervert the work that Meitner had pioneered on nuclear fission before it could be used for good. Meitner was a highly ethical scientist and the knowledge that her scientific work was the principle behind the first nuclear weapon, developed during The Manhattan Project (which she actively avoided having any involvement with), was clearly on her conscience as she spent some of her retirement giving talks about the social and moral responsibility of scientists.

Fortunately, Meitner was still alive when fission was put to the less destructive use that went on to affect the world we live in today. The first nuclear power plant was built in 1951.[11] Although the figures fluctuate for a variety of complex reasons, as of 2019, around 11 per cent of global electricity was provided by nuclear power.[12] Some countries are reluctant to use nuclear power due to safety concerns raised by historical incidents, whereas some fully embrace it, with France generating around 70 per cent of its electricity from nuclear sources in 2019.[13] Lise Meitner's dedication to science and resilience in the face of one of the most extreme forms of anti-Semitism, provided future generations with a glimmer of hope in the continuous and vital challenge to promote the use of energy sources that reduce carbon emissions.

Rufaida Al-Aslamia (Born 620)
Nursing
Persia

Nursing has historically been one of the disciplines considered an 'acceptable' profession for women. Mary Seacole was a pioneering nurse whose compassion and talent for healing improved the lives of soldiers injured in the Crimean War, and Florence Nightingale became known as the founder of modern nursing due to her introduction of improved sanitation, visual statistics, and contributions towards healthcare reform during the same conflict.

As inspirational as these nurses were, there were many other interesting women in healthcare who preceded them. One such woman, Rufaida Al-Aslamia, was considered to be not only a social worker but the first female Muslim surgeon and nurse.[1] She was born around 620 AD to a father who was a medic. The early days of Islam enabled women to access opportunities for learning[2] and Al-Aslamia did so by assisting her father with his work. She was active in her profession during the time of the Prophet Muhammad; her name and work was documented in Sunnah, the collection of customs and practices that guide the Islamic community. She had a hands-on role during the holy wars, tending to wounded and dying soldiers,[3] but was also an important figure in the community; having been given permission to set up a tent next to a mosque by the prophet himself, Al-Aslamia looked after patients who were in need of medical advice.[4] In the pre-Islamic period, the provision of healthcare had been basic in terms of patient contact time. The physician would conduct home visits with the sole purpose of diagnosing and prescribing medication.[5] Al-Aslamia revolutionised healthcare by showing the benefit of adding an element of emotional care, comfort, and support. She taught this novel concept of nursing to other women and their numbers within the community grew steadily. The influence she had on her field has not been forgotten to this day; the University of Bahrain commends exceptional skill in nursing with the annual Rufaida Al-Aslamia Prize, and several educational institutions across the Middle East have been named in her honour. Faith would have been an important part of her life since accepting Islam, and this guided her morals. Another vital legacy she provided for future healthcare workers was a code of ethics; a concept that has been intrinsically linked to the benefit of patients worldwide ever since.

Chapter 2

Our Day-to-Day Lives

Unbeknownst to the majority of people, many things we have come to take for granted as part of privileged, twenty-first-century living had a female mind behind them. Given the historically traditional female role of homemaker, carer for children, and completer of household chores, it should not be surprising that over the centuries, women have come up with countless devices to save time, effort, and money. We are curious beings at heart and, when struggling to balance tasks or overcome problems, it is only natural to try and find a solution to improve the situation. Process improvement and workflow is a vital part of most engineering and technology disciplines so despite many STEM fields being stereotypically male-dominated, there are many women who used, and continue to use, their gender-typical upbringing to their advantage rather than being restricted by it. Whilst most female-led innovation started in the home, legal and social improvements gradually allowed women to prove that whether working in an office, laboratory, or agricultural setting, they are just as capable of being creative and improving the world around them as their male counterparts.

This being said, womankind's relationship with patents has been rather disappointing. All over the world, the legal rights of married women are likely to have been a huge deterrent, if not barrier, to the pursuit of invention. Writing in 1883, Matilda Joslyn Gage referred specifically to the plight of women living in the US. She explained that even if a woman were to successfully obtain a patent, she would not hold the title, or the rights to the contract, nor the ability to licence the use of her invention – and in the event of her patent being infringed, she would struggle to take legal action.[1] In the story of female inventors, social and economic factors have had a role to play, as well as setbacks surrounding the legal status of women. For many of us who were fortunate enough to enjoy a good education as we grew up, it is difficult to imagine a world in which women were point-blank denied the opportunity to learn and to put that knowledge to use to earn a living. Yet the idea of a career or technical education has most certainly not been an option for the female sex for most of human history. As is so often the case, certain events allowed some progress to be made. Major conflicts such as the American Civil War and the First and Second world wars provided the opportunity for a huge social shift in the day-to-day roles of women, often facilitating them with better education and access to skills outside of the home. As property rights in several countries started to change, and expositions and exhibitions were used to showcase the capabilities of female inventors, there will have been a notable increase in patent activity. Despite this progress, in the Western world, barriers continued to stay in place for most women who did not come from a white, middle-class background. As several of the following inspirational biographies highlight, it is vital that we aim to grow diversity in more than one direction; that if we are truly to pull together to create a more equal, sustainable world, we need to be more open-minded than focusing on one single aspect of making the world a better place. It is also telling that most of the inventors presented in this chapter lived in the US; information about female inventors who worked in Asia and the Middle East is far more scarce. It may well be that social structures and patent application processes differ, or simply that biographies originating from these places have not been translated from the native language and are therefore more challenging for Western readers to find. Nevertheless, the statistics are certainly encouraging. If we momentarily look past the fact that the report published by the Intellectual Property Office in 2016 states that only 7 per cent of inventors around the world are female,[2] it is clear to see that the number of inventions is increasing. In terms of sole inventorship, the 2019 version of the same report shows that the proportion of females increased from just over 2 per cent in 1915, to 12.7 per cent in 2017.[3] The data on worldwide female participation in patent applications shows an even larger increase, from around 2.5 per cent in 1915, to over 20 per cent in 2017.[4]

However, as the lives of the following women show, there are many different ways to change the day-to-day lives of people all over the world. Although they worked in a wide range of environments, these ladies all had one thing in common: the ability to spot the need for improvement, and the motivation to take a chance on something not yet attempted.

Alice Parker (1885–?)
Design and Technology
United States

For the woman whose patent was the premise behind the central heating that keeps many homes warm in cold weather, we know shamefully little about Alice Parker. A quick internet search will bring up the photographs of two completely different women, neither of which come from a source that can confirm whether the woman depicted is in fact Alice Parker! Census records show that she was probably born in Virginia around 1885 to African American parents[1] and attended a high school affiliated with Howard University. Beyond this fact, no information regarding her education has come to light. The 1920 census states that by this time (just a year after she received her patent), she was working as a cook for a family in New Jersey and that she was married to a man who was a butler for the same household.[2]

Around the time Parker received the patent for her gas furnace, heating American houses was time-consuming and inefficient. Wood had to be cut and stored correctly and even once logs were placed in the fireplace, heating rooms evenly was a challenge. The air ducts she designed allowed heat to reach different areas of the building from a single furnace, but her idea had additional benefits. The patent outlines the aims as saving labour and fuel, as well as simplifying the operation of heating the building. This was achieved because of the ability to control the different units within the furnace, so heat could be independently regulated or shut off in rooms that were not in use.

Although Parker's gas furnace patent went some way to reducing fire risk as fires would no longer need to be kept lit overnight, there were new safety concerns which ultimately proved to be a barrier to the implementation of the system. Nevertheless, this innovative woman who was a pioneer of thermostat-controlled heating systems has not been forgotten and her name lives on in the Alice H. Parker Women Leaders in Innovation Awards, which is distributed by the New Jersey Chamber of Commerce.

Josephine Cochrane (1839–1913)
Design and Technology
United States

In 1850, Joel Houghton was the first recorded person to register a hand-cranked dishwashing device.[1] It remained unsuccessful in commercial terms, so L. A. Alexander submitted a patent for a similar machine. It too was unsuccessful.

Bring in Josephine Cochrane (née Garis), who certainly had more going for her than simply being a rich, pretty socialite! As a wealthy couple residing in Shelbyville, Illinois, Cochrane and her husband, William, were no strangers to entertaining and the number of dirty dishes surely reflected that. Unfortunately, it seemed that Cochrane and her kitchen staff were in some disagreement about the care with which the household china was handled, and as such it is said that the lady of the house turned her hand to washing up. Perhaps unsurprisingly, this arrangement did not last as she soon found the task to be time-consuming and tedious.

Cochrane, who was born in Ohio in 1839,[2] had a father who was a civil engineer by profession and a great-grandfather who had been an inventor, so she may well have had a natural affinity for innovation, despite not receiving any technical education.

Whilst working on the design of her machine, personal tragedy struck as William passed away. This may have made her more determined than ever; Cochrane was not content to stop working even after she received her patent in December 1886.[3] Her entrepreneurial mind started to shine as she went on to set up the Garis-Cochrane Manufacturing Company to produce her design. No longer satisfied with making dishwashers for her friends and advertising within her own social circles, she set her sights on widescale commercial usage. She hired a man called George Butters to help build her prototype and he was the first employee at her new company. By the time the 1893 World's Columbian Exposition in Chicago came around, Cochrane had managed to get nine dishwashers installed in restaurants that were catering for the event; by doing so, she was able to reach a larger audience and attract the attention of other restaurant and hotel owners.

For the rest of Cochrane's life, the machine she invented mainly served commercial environments as internally plumbed hot water was not available in many homes. However, after her death on 3 August 1913,[4] the Garis-

Cochrane Manufacturing Company was acquired by KitchenAid,[5] and a combination of modern plumbing and the shift in attitude regarding more traditional domestic arrangements brought Cochrane's invention into the next century. From 1949 onwards, KitchenAid started to manufacture dishwashers based on her design.

With health and safety standards of industrial food preparation areas developing over the last century, dishwashers have become a necessity for ensuring that dinnerware is sanitised – so spare a thought for Cochrane the next time you eat something that was prepared in a shop or restaurant. And with more modern dishwasher designs incorporating energy and water-saving measures, those of us who have dishwashers in our homes may be using them and reducing our environmental impact without even being aware. An impressive result from what started as a disagreement over broken china.

Further Reading

Marie Van Brittan Brown – an African American nurse who invented a home security system that included cameras, audio communication and a button to alert law enforcement and emergency responders in case the intruder proved a threat to inhabitants. She filed a patent (US3482037A) for this invention with her husband in 1966.

Mary Anderson (1866–1953)
Design and Technology
United States

Picture if you can, riding on a streetcar in New York. The year is 1903 and it's snowing. Apart from people's clothes, buildings and the lack of mobile phones in the hands of your fellow passengers, there is one other fairly big difference that draws your attention. Every so often, the driver stops, gets off the streetcar and wipes snow off the windscreen with his hands.

Evidently, the delayed journeys caused by this procedure didn't bother most people at the time, but it didn't go unnoticed by the innovative mind of real estate developer, Mary Anderson.[1] She was so captivated by seeing this during her trip to New York that when she returned home, she hired a designer and came up with a device with rubber blades that could 'wipe' the windscreen, whilst being controlled from inside a vehicle. A windscreen wiper to you and me.

Anderson filed for a patent during the summer of the same year and it was granted on 10 November 1903. Unfortunately, her solution appeared to be ahead of the times; she was told by one company to which she attempted to sell the rights that it was not deemed to have any commercial value. This seems amusing now, as the windscreen wiper is such a standard feature of all cars. However, it wasn't until Henry Ford started mass-producing cars that vehicles became more affordable; Anderson had the misfortune of creating her invention before this industrial turning point. Cadillac was the first company to include her design as standard; although it must have been gratifying for her to see her invention have such widespread use, the injustice of never receiving any money for her efforts may have been frustrating. Fortunately, Mary Anderson has not been forgotten and continues to receive full credit for the way she improved transport safety in the modern world.

Mary Beatrice Davidson Kenner (1912–2006)
Design and Technology
United States

Mary Kenner (née Davidson) filed five patents in the space of thirty-one years,[1] and should be remembered for the pragmatic yet creative innovator that she was.

Kenner had a dramatic start to life. At 5 years old, her attempt to help out in the garden (to improve her chances of receiving a visit from Santa Claus) had ended in disaster, when she unknowingly picked up an armful of dry leaves that had been set alight.[2] Although her mother was quick to act and all five of the town doctors were called, they informed her parents that the burns were too severe and she was going to die.[3] Fortunately, with a loving family to watch over her day and night for six weeks, Kenner survived and regained her health.

Born into a family of inventive minds on 17 May 1912, Kenner was sure to have been influenced by stories of her grandfather's tricolour train signal and her father's sadly unsuccessful venture into inventing a portable trouser press. Looking back during an interview, Kenner recalled that the first time she thought of an invention was only a year after her accident – an impressive achievement for a little girl who had been almost left for dead. Despite having facial disfigurations from the burns she had received (their local hospital hadn't received the necessary training to treat burns) and the trauma that came with it, the 6-year-old Kenner focused her attention on the squeaky door of their family home. She noticed that the door always seemed to revert to making a noise when opened, no matter how many times she watched her father oil the hinges. She very sensibly suggested that someone should invent a self-oiling door,[4] and took herself off to do just that. Being 6 proved to be a little too much of a setback, so the project was dropped.

However, the ideas continued to come to her. When her family moved from Charlotte, Carolina to Washington DC, the now 12-year old Kenner spent time wandering around the United States Patent and Trademark Office to check whether anyone else had registered similar devices to the ones she had dreamed up. By the time she reached adulthood, she had come up with the idea that made her a pioneer of women's hygiene products. Her idea for a sanitary belt involved a strap that tied around the waist and held in place a waterproof liner into which sanitary pads could be inserted. Kenner recalls receiving interest from the large brand Johnson & Johnson, but it was a business called the Sonn-Nap-Pack Company who were curious enough to send representatives to come and meet the inventor of this novel product. Upon seeing a young black woman answer the door, the company representatives were apparently very quick to withdraw their interest.[5] One can only imagine how disappointed Kenner must have been by this injustice; she admits to having thoughts about houses and cars and other things she would be able to do if she could sell the rights to her design. She must have continued to follow the activities of this company as she became aware that they attempted to patent her design after rejecting her. Fortunately, they were unsuccessful but other companies did start to manufacture Kenner's design after her patent expired. She received no credit or money.

Although many women inventors who had come before her relied on selling their inventions to make money, it seems that this was never Kenner's primary aim. Patents were expensive to obtain, and she paid between $200 and $2,200 for each of the five she applied for.[6] Nevertheless, Kenner took great joy in noticing inefficiencies in the world around her and coming up with novel ideas to improve them and make life easier for people. It is estimated that she came up with over 100 inventions,[7] ranging from a tray that could be attached to a walking frame to portable garages and accessories for cars.

When not inventing, Kenner worked for the Census Bureau and the General Accounting Office but decided to leave when it was clear that she was being passed up for promotion due to racial prejudice.[8] She then went on to run a number of flower shops successfully for over two decades, and later fostered five children with her husband.

Stephanie Kwolek (1923–2004)
Chemistry
United States

Stephanie Kwolek was a chemist whose bullet-stopping innovation led to thousands of lives being saved. Born in Pennsylvania on 31 July 1923, Kwolek grew up with an interest in both fashion design and science, inspired by her mother and father respectively. Her father passed away when she was only 10 years old, but the time they had spent together exploring nature was clearly influential as Kwolek chose to study chemistry at university. She was accepted into Margaret Morrison Carnegie College of Carnegie Mellon University and graduated with a bachelor's degree in 1946.

Due to the requirement for men to join the army and serve overseas, several job vacancies had opened up at large companies and Kwolek decided to apply for a job with the chemical company DuPont, to save enough money to attend medical school – her dream at this time. Kwolek enjoyed her work there but, little did she know, a career-defining moment was just around the corner. Her role was to work on processes to synthesise high-strength petroleum-based fibres that could function in extreme conditions.[1] Polymers are made up of repeating units, and Kwolek would spend her days dissolving polymers with a specific type of repeating unit in a solvent, then using this solution to spin fibres. She was also interested in the conditions under which the process was carried out; factors such as temperature can have huge effects on the properties of the fibres produced. Usually, the solution used to spin the fibres was clear and fairly viscous. One day, under a specific set of conditions, Kwolek found that the solution had turned cloudy and was very fluid. Whilst others may have dismissed this deviation as a mistake, thrown away the liquid and started again, Kwolek decided to stick with it. It may seem like an easy decision to make in hindsight; however, laboratory equipment is not cheap and there may well have been severe consequences if the spinneret had been damaged as a result. When this new fibre was tested to the strength at which nylon would usually break, it did not. Kwolek had created something called a liquid crystalline solution; once spun, this meant that the molecules within the fibre were highly orientated, making it incredibly strong. DuPont was quick to realise the significance of this discovery, and a team worked on developing this new fibre for use in the real world. By 1971, this fibre had been branded as Kevlar® and introduced to the world. Due to its aligned and tightly spun microstructure, this material has properties that dissipate the energy of high-velocity projectiles and provide a barrier against slashes and punctures. Kevlar® quickly became a material associated with protecting service people, police officers, and workers operating in dangerous environments. According to a partnership between DuPont and the International Association of Chiefs of Police, 3,100 lives were saved in the thirty years after the IACP/DuPont™ Kevlar® Survivors Club® was founded in 1987.[2] For a material that was developed from what was thought to be waste product, Kwolek's perseverance paid dividends.

Further Reading

Elizabeth Fulhame – a chemist who discovered the vital chemical reaction that is now known as catalysis. In a book published in 1794, she recorded details of the experiments that led to this ground-breaking discovery. She also experimented with light-induced reactions that would eventually pave the way for the development of photography.

Ruby Hirose (1904–1960)
Biochemistry
United States

Born in Washington in 1904[1] to Japanese parents, Ruby Hirose showed the society of her time that her status as a Nissei (second generation immigrant) was not a barrier to doing valuable scientific work that benefited people all over the world. Hirose's parents had moved from Japan to the US in search of a new life when their manufacturing business failed,[2] and took up farming on land they leased in the White River area of Washington. This is where Hirose grew up and, despite having no experience of discrimination during her own childhood when she was the only student with Japanese heritage in her school, she saw the gradual development of anti-Japanese sentiment as her younger siblings progressed through their education.[3] The demographic of the US was becoming more diverse; plentiful work created by the building of new infrastructure had attracted foreign workers, many of whom settled and had families. Although a section of the Naturalisation Act of 1870 had allowed people of African descent to be included in the naturalisation process, this did not extend to immigrants from other countries.[4] Because of this, the Hiroses were not able to own land unless it was bought in the name of their children who had been born in the US. The family's hardship only got worse, as Hirose's mother and younger sister, Fumiko, contracted tuberculosis. By the time Hirose had moved away to study pharmacy at the University of Washington, they were both confined in Firlands Hospital.[5] Fumiko died whilst her elder sister was completing her degree, and their mother died less than ten years later.

After receiving her master's degree, Hirose relocated to the University of Cincinnati in Ohio, where she researched a coagulation factor called thrombin. In her thesis, part of which was published in the *American Journal of Physiology* in 1934,[6] she notes that the mechanism by which thrombin action works had recently been re-evaluated. Among other functions, thrombin acts to convert soluble fibrinogen to insoluble fibrin, which helps the formation of blood clots. It had been proposed that this formation of fibrin did not signify the end of thrombin action as previously thought, and Hirose studied the process in order to provide further evidence.[7] By now, she had shown herself to be a talented researcher and in 1931, Hirose was one of two people to be awarded the Moos Fellowship in Internal Medicine.[8]

Having earned her PhD in the early 1930s, Hirose was then employed by William S. Merrell Laboratories, initially to conduct research on serums and toxins. The work she did had an important impact on global health, especially for children, as it laid the foundations for the Polio vaccine. Polio is a disease that can have long-term symptoms including paralysis, muscle weakness and deformity. Once widespread, cases have decreased by over 99 per cent since 1988 due to an extensive vaccination programme.[9]

Another disease Hirose studied was diphtheria, a contagious bacterial infection that can be fatal. In the early 1930s, it had been suggested that the toxin could be precipitated using potassium aluminium sulphate, commonly known as alum.[10] Hirose continued this research, which contributed to the development of the diphtheria vaccine.[11] Using the same logic, she then found that alum could also be used to treat the pollen extracts that were the basis of hay fever treatment at the time.[12] These extracts were used to desensitise people with allergies. By increasing the effectiveness by using alum,[13] Hirose managed to refine the treatment.

In 1942, President Franklin Roosevelt issued Executive Order 9066, which authorised certain areas of the country to be treated as military zones and led to the incarceration of people of Japanese, Italian, and German descent.[14] These citizens who lived in the Pacific Northwest and Westcoast regions were deemed to be a threat purely because of their ethnicity and were forcibly evacuated to internment camps. Two-thirds of these people, including Hirose's brother and sister, were American-born Nissei with little experience of Japanese politics or culture.[15]

During this time, in which Hirose must have felt huge concern about the well-being of her family, she continued with her research in Ohio, where she was safe from the threat of incarceration. More than ever she may have experienced the identity crisis of belonging to neither the society she was born and bred into nor the Japanese society of the first generation, or Issei. This sentiment was vocalised in a booklet written by the Japanese Christian Students Association, an organisation she had joined during her time in Washington. Fortunately, her local scientific community did not feel the need to discriminate and she became one of ten female members of the American Chemical Society Cincinnati branch.[16]

In later life, Hirose specialised in bacteriology, working in several different hospitals. She developed leukaemia and died at the age of 56 in 1960,[17] the same year that the Treaty of Mutual Cooperation and Security between the United States and Japan was amended in an attempt to improve relations between the two countries.

Olive Dennis (1885–1957)
Engineering
United States

The childhood of Olive Dennis, the world's first railway service engineer, would have been extremely different from that of her peers. She was interested in building, even more so after her father gifted her a tool kit in an attempt to stop her from using his,[1] and took an interest in the construction sites on her way home from school. Having been born in 1885, a traditional career choice was to go into teaching – a job she had for ten years after obtaining her two degrees, a Bachelor of Arts and a Master's in mathematics.[2] However, passion does not always dissipate with time and Dennis found this to be the case with her long-standing interest in civil engineering. Following her heart, she enrolled at Cornell University and managed to obtain her degree in just one year.[3] When she graduated in 1920, she was one of the first females to complete the course. Dennis managed to gain employment the same year and started working as a bridge design draftsman for the Baltimore and Ohio (B&O) Railroad.

It was a challenging time for the railway industry. Henry Ford's work on car production processes and the introduction of the assembly line[4] had made motor travel affordable for the middle classes and this inevitably had a knock-on effect for existing modes of transport. This may have been one of the factors behind the market research initiated by Daniel Willard, president of the B&O Railroad. He made a suggestion that as an estimated 50 per cent of passengers were female, more effort should be made by the company to secure their ongoing custom.[5] He also felt that comfort upgrades aimed at women should be overseen by a female engineer, so the novel position of 'service engineer' was created and offered to Dennis. And she did not disappoint; her innovations were plentiful and significant. Small luxuries such as reclining seats, liquid soap in bathrooms, and air-conditioned carriages are things now common in train interiors, but it was Dennis who took the initiative to introduce these things to the transport industry. Long haul flights would be much less comfortable without the dimmable lighting that Dennis originally designed for railway carriages. Throughout her career, she is said to have travelled around 50,000 miles a year,[6] testing carriage environments so the company could ensure customer experience was as good as it could be. And although she was responsible for more creative aspects of design, such as waterproof upholstery and colour schemes of dining cars, it appears she was far more interested in the blueprints, layouts, and mechanical workings of carriages.[7] One of the most significant inventions she received a patent for was a ventilator that allowed passengers to get fresh air without being exposed to draughts.

Perhaps the highlight of her career was being given an assignment to redesign four P-7 'President' class Pacific locomotives. All of her previous contributions were incorporated into the Cincinnatian; Dennis also went as far as to streamline the front of the locomotive to make it more accessible for mechanics. From the information provided in her patents, it is clear that Dennis was creative, thoughtful, and showed great attention to detail.

As is common, many of her patents were registered to B&O rather than to Dennis herself, so her name is not as well known as it should be. Nevertheless, she was at least recognised for her work at the time as she became the first female member of the American Railway Engineering Association.[8] In summary, Dennis showed just how necessary it is to have a diverse workforce in order to solve problems. 'Traditional' roles sometimes need questioning in order to make progress.

Further Reading

Beatrice Shilling – a British aeronautical engineer and racing driver whose development of a fuel restrictor solved the issue of Rolls-Royce Merlin engines becoming flooded and causing malfunction in Hawker Hurricane and Supermarine Spitfire aircraft.

Dr Barbara Sabey – a British road safety engineer who is best known for her work at the Transport Research Laboratory. She was involved with the implementation of seat belt laws, tyre tread depth requirements, and legislation on drink-driving.

Sutayta Al-Mahāmali (920–987)
Mathematics, Law
Modern-day Iraq

The history of science and maths may often be viewed through a Western-centric lens; however, there were scholars and great centres of learning in the Middle East long before the likes of Isaac Newton came onto the scene. Muhammad ibn Musa al-Khwārizmī was born in Persia in the eighth century and published the first text on algebra[1] at a time when European countries were more interested in carving up territory and waging wars on each other. He is often called the 'father of algebra' and was so ingrained in the history of mathematics that his name 'al-Khwārizmī', meaning 'native of of Kwārizm', became one of the origins of the word 'algorithm'. [2] He was also in charge of the library at the House of Wisdom in Baghdad.[3]

More than a century later, it was a woman who was causing ripples in the intellectual circles of the same city. Born around 920,[4] Sutayta Al-Mahāmali was a mathematician who also had a talent for law. Her knowledge was widely respected and being female did not seem to cause her any issues; an unusual privilege in terms of women's history. Her particular mathematical interest was algebra and like al-Khwārizmī she was good at developing ways to find general solutions to problems. Practically speaking, algebra was just as useful in tenth-century Islamic society as it is today. Dividing up inheritance according to Islamic Law was extremely complicated, but by using mathematics as a tool, Al-Mahāmali was able to provide fair solutions for the families who came to her for advice. And it wasn't just families who wanted help; scholars would travel to see her when they were stuck on a mathematical problem.[5] Even more impressive is that her work and original contributions were consequential enough to be cited in later mathematical texts.[6] She was truly a visionary; it would be hundreds of years before fully educated women of equal standing would be documented in European history books.

Al-Mahāmali showed the kind of impact that could be had by developing a good understanding of multiple fields and not being content to specialise in one subject. For most of recorded academic history, learning has been broad and it was the norm for mathematicians and scientists to engage in a range of subjects including the arts. Al-Mahāmali embraced the opportunities she was given and used her expertise to help others, ultimately providing an example of how individuals and society as a whole can benefit when girls receive a good education.

Chapter 3

Improving Lives, Equality, and Justice

Margaret Sanger (1879–1966)
Nursing, Social Reform
United States

Throughout the course of human history, pregnancy and childbirth have posed a very real threat to female health and livelihoods. Margaret Sanger (née Higgins) was one of the pioneers who recognised the importance of women being able to have some control over if and when they have children.

Sanger's own mother had conceived eighteen times in twenty-two years and, as a nurse, she saw plenty of heartbreaking situations in which women had died from attempting self-induced abortions. Although unwanted pregnancies still hold a certain stigma today especially amongst the teenage population, it is no longer a life-or-death issue in countries that provide access to birth control and reproductive health services.

Sanger was a great advocate for teaching sex education to teenagers and held the belief that doing so would prevent the act of sex from being so glamorous to men, as well as helping women to better understand their bodies. In one of the many articles she wrote for the *New York Call*, Sanger notes how the passage of girlhood to womanhood is often ritualistic within tribes; although her use of language is very dated, her praise of information regarding sexual relations being made accessible to children as they reach puberty shows that her views were well ahead of her time. Sanger had a great understanding of teenage development and clearly felt that relational and social problems were commonly caused by a lack of information available to children as they approached adulthood. Believing that 'civilization [was] still hiding itself under the black pall of prudery',[1] Sanger wrote a series of articles called, 'What Every Girl Should Know'. These explained sexual anatomy and sexually transmitted diseases in detail, in an attempt to provide young people with a source of information they could easily access. As with many pioneers, Sanger's drive to educate, or more specifically her topics of choice, were not well received by more conservative sections of society. It is easy to imagine how horrified some readers would be, knowing that the adolescents they were trying so hard to keep pure of mind and spirit had access to writings that graphically outlined subjects such as sexual desire and masturbation. Sanger went on to publish and circulate a monthly newsletter, *The Woman Rebel*, to promote contraception. This time, she not only received further bad press but also faced trial for violating postal obscenity laws under the Comstock Act.

Sanger crossed the Atlantic to avoid trial but during her time in Europe had many experiences that would further her cause; to allow women to choose whether or not to conceive. Time spent in a Dutch birth control clinic gave her ideas that she would implement back in the US when she co-founded the first family planning clinic of its kind in Brooklyn in 1916. The clinic was immediately popular, but it only took nine days for the police to find it and enforce its closure.[2] This time, Sanger was found guilty of distributing birth control information and sentenced to a month in prison. By this time she had children so her activism continued at great personal risk to her family. After being released, she continued to advance her campaign, setting up the American Birth Control League and the Birth Control Clinical Research Bureau in the same year. By establishing these organisations, she hoped to bring structure to the goal of helping women to control their own fertility, as well as gathering data on the long-term safety and effectiveness of birth control. Her travels abroad only added to her conviction that widespread information and clinical services were the solution to preventing 'enforced motherhood'. She held a low view of abortion and was particularly horrified by infanticide, which she had come across whilst travelling to observe reproductive health in other cultures.

The birth control movement had well and truly begun; however a turning point was the opening of several more clinics. She was invited to open a clinic in Harlem, New York in 1929 by the leader of the city's Urban League, James H. Hubert.[3] She staffed the clinic with black doctors and nurses and it was overseen by a board of prominent members of the black community in New York.[4] This included W.E.B. Du Bois, who went on to found the National Association for the Advancement of Colored People (NAACP).[5] During her stint here she was reported to have been tough on members of her team who exhibited racial intolerance.[6] Her efforts to work alongside these communities to provide family planning services and improve quality of life led to her being praised by Martin Luther King Jr.[7]

Following the court case 'United States v. One Package of Japanese Pessaries' which was triggered when a doctor ordered a diaphragm from Japan, the law was changed to permit the use of contraceptives in the event that pregnancy would cause harm to the mother or child. Further success came in 1936 when contraceptives became legalised in New York, Connecticut, and Vermont. Just a year later, the American Medical Association showed its support for birth control devices being standard medical practice. It would take decades for the rest of the country to follow suit, but it was a pivotal moment as it showed that public opinion was starting to change.

The life of Margaret Sanger cannot be discussed without addressing her involvement with eugenics. On the surface, it may seem obvious that campaigns to publicise the use of birth control were born out of the women's movement. However, multiple sources suggest that it was also used as a more sinister tool for eugenic agendas.[8] Ever since this concept was given a name by Francis Galton, cousin of Charles Darwin, it has been a complex and controversial issue. In the early 1900s, there was support for Galton's idea of artificially harnessing his cousin's theory of natural selection to selectively breed in order to eradicate 'undesirable' qualities and pass on good ones to future generations. The US implemented eugenics policies around this time,[9] and the idea also gained popularity in France, Germany, and Great Britain.[10] The Nazi Party was infamously pro-eugenics and the actions they took were a devastating example of how a concept that may have started as an innocent solution to 'improve' the human race was in reality extremely cruel and unethical. During the interwar years, eugenics and birth control could be described as almost interchangeable,[11] so it is easy to see how, as someone so involved with the birth control movement, Sanger's initial aims might have been achieved at the cost of other humanitarian causes. Since moving to New York with her first husband, Sanger had been an active socialist; her book *Woman and the New Race* shows her belief that socialism, feminism, and birth control were intertwined. Like the early proponents of eugenics, Sanger expresses passionately in her writings that uncontrolled population growth was the cause of everything from poverty to child labour. Angela Franks, author of *Margaret Sanger's Eugenic Legacy: The Control of Female Fertility*, summarises Sanger's world view as 'quality, not quantity'.[12] Whilst this opinion may well come from an earnest desire to reduce suffering in the world, it is overly simplistic. It does not take into account the ethical question of who gets to decide what is classed as 'desirable' and suggests that people at risk of being born into suffering, disease, or poverty should not be born at all. It goes without saying that this is an inherently problematic point of view.

Whilst Sanger frequently expressed her view that lack of education amongst the lower classes was partly responsible for the poor quality of life that comes from overpopulation, documentation regarding her interaction with black and ethnic communities can appear contradictory and has led to much controversy. Whilst working on the Negro Project, an initiative set up to increase access to birth control in the Southern states, she sent an ambiguously worded letter advising that a greater number of black ministers should be brought on board.[13] It was clear that Sanger wanted to avoid accusations that her work with black populations was in any way racially driven but it has since raised questions of her intentions.[14] In the end, the project was not as successful as it could have been. Sanger's plan involved a year of education before rolling out clinical services, and she placed particular importance on clinics being run by black medical professionals, given many communities' experience of racially fuelled mistreatment and abuse that stemmed from white supremacy. The original plans were scrapped against Sanger's will and the initiative was instead set up in existing white-run clinics. In contrast, in her own autobiography Sanger talks of a meeting with the women's branch of white supremacy group, the Ku Klux Klan. Her willingness to associate with such an openly racist organisation that could easily use her campaign to advance their own agenda has been heavily criticised.

Ultimately, whilst Sanger's evidently genuine aim to reduce the suffering of women caused by unplanned pregnancies was a great step towards female emancipation, any positive effect is lost as soon as the decision to conceive is taken away from the individual woman. The controversy surrounding Sanger undeniably stems from how far she went with linking the reproductive autonomy of individuals to the ideology of population control and discouraging 'unfit'[15] people (usually those from lower economic backgrounds) to procreate. This desire to exercise control over others is certainly problematic, and it cannot be ignored that a prestigious media award named after her has been rebranded in recent years. Even Planned Parenthood, the organisation she had so much involvement with has found it pertinent to consider changing the name of its building.[16] It is often challenging to look at the views of respected historical figures, given that the concept of 'acceptability' shifts continuously over time. In some cases, it is possible to view scientific developments in an abstract way and not allow the sole focus to be on any negative use of the technological advancement. After all, the pioneer whose work contributed to it may not be aware of the intended use. However, history has shown that some pioneers have used their work to intentionally harm others and it is evident that others may do so in the future. This can never be ignored, as such individuals have the power to inflict untold horrors on other human beings.

Sanger has certainly had a lasting effect on our world, although exactly where the dial falls between positive and negative is an ongoing topic of discussion. Her early days of being persecuted by the authorities served as inspiration for the creation of the superhero 'Wonder Woman', though it was a closely guarded secret due to the comic creator's unorthodox relationship with Sanger's niece.[17] Thanks to her tenacity and organisational skill, Sanger provided an incredible service to the health of women by putting herself at substantial risk to publicise issues regarding reproductive rights. Although there are conflicting opinions on whether birth control is ethical, the ability for women to decide whether or not they become pregnant is an undeniably powerful thing that has changed the lives of so many. However, evidence does point towards the fact that her beliefs around the implementation of this development were imperfect and, ironically, may have caused a lot of harm to the very women she had embarked on her career to help.

Further Reading

Helen Rodríguez Trías – a Puerto Rican paediatrician who campaigned against sterilisation abuse and became the first Hispanic president of the American Public Health Association.

Clelia Duel Mosher – an American doctor and women's health advocate, whose extensive research challenged the existing opinion that women have no sexual desires. She also worked on improving menstrual hygiene and showed that the perceived physical weakness of women was largely due to the societal expectation for them to wear corsets, which restricted breathing.

Anne McLaren (1927–2007)
Developmental Biology, Genetics
England

A dame commander of the British Empire, the first female officer of the Royal Society,[1] and a valued member of multiple ethics committees, Anne McLaren was a biologist whose pioneering work contributed to the possibility of *in vitro* fertilisation (IVF).

McLaren was born in London in 1927, to Sir Henry Duncan McLaren, 2nd Baron Aberconway, and Christabel Mary Melville MacNaghten. When the Second World War broke out, the family moved to Wales, though McLaren completed her schooling in Cambridge. As a bright young woman, it appeared that she could have chosen from a range of disciplines to study, but the biology entrance exam for Oxford University looked easier than the daunting reading list given to those wishing to apply for English Literature.[2] And so by complete chance, McLaren started on a career that would provide hope for couples who require assistance to conceive and a guiding hand for societies that had to build and navigate the legal and ethical framework for a world in which IVF had become possible.

McLaren's university experience was affected by the Second World War since the conflict had resulted in a new way of thinking for many, with some traditional social norms being left behind. Some of the lectures she attended were influenced by the tutors' contributions to the war effort and, through the access she received to study a range of courses, McLaren developed a keen interest in genetics. Graduating with an excellent grade, McLaren decided to continue her studies. During these years, she studied genetics in rabbits and viruses in mice.[3] After completing her PhD in zoology in 1952, McLaren chose to specialise in genetics and found a research position at University College London (UCL) alongside Donald Michie, her fellow student and soon-to-be husband. Together they conducted novel research in transferring embryos between two strains of mice and investigating whether the maternal effect was caused by the uterus in which the egg was implanted or the egg itself. The couple had a very pragmatic approach to their work, spending some money from a previous grant on baking tins from Woolworths so they could build a bespoke mouse house. This new accommodation for their test subjects was so successful that the number of mice soon required more space; in 1955, their work was moved over to the Royal Veterinary College. Their most significant work was yet to come. Working with John Biggers, a researcher who was creating cell culture in the laboratory next door, McLaren was able to show that embryos that had been stored outside of the mouse for twenty-four hours could be transferred into the uterus, grow and be born. The mice bred using this method of fertilisation even survived until adulthood. Within a couple of decades, this same method would be successfully tested in humans.

With this hugely significant work under her belt, McLaren continued her research, moving to Edinburgh in 1959 to join the Institute of Animal Genetics, where she started her work on chimera. A chimera is an organism that contains two sets of DNA, and these are what McLaren used to study environmental and inherited factors. Both of these factors can affect the way in which embryos develop, and insight into them provides an understanding of complications that may arise.[4] She was one of the earliest pioneers to recognise what this method of research could be used for. McLaren's thoughtful and careful design of experiments was just as impressive as the results she achieved. Because of this, she was honoured with an award from the Zoological Society. The way in which she controlled factors during her studies provided 'unambiguous answers', (in science, repeatability is key), and McLaren's methods were clear enough to be replicated by colleagues all over the world.

Years later in 1974, she returned to UCL to take on the directorship of the newly established Mammalian Development Unit of the Medical Research Council.[5] Back in London, McLaren and her team turned their attention to building on the embryo research she had done with John Biggers and her now ex-husband, Michie. Her ambition of producing research that had a practical outcome was realised in the years that followed. Experimentation on embryonic development meant that embryos could be genetically profiled before being implanted and this allowed the screening of serious genetic diseases. This made its way into a Parliamentary Bill in 1990 and it passed, likely due to the success of the first clinical trial.[6]

Despite this achievement, it is clear that McLaren understood the implications of the work she was doing and that these could be positive and negative in equal measure. Altering the genetic make-up of organisms and manipulating cells is close to what could be described as 'playing God'. In 1982, McLaren became the sole biologist on the Warnock Committee[7] set up by the UK government to research the ethical, legal, and social issues surrounding embryology and human-assisted fertilisation. She and her colleagues did important work by providing recommendations. One particularly significant example is that it is vital for IVF to be regulated by an authority. As well as setting important ethical boundaries, many of the points put forward by the Warnock Committee were empathetic and carefully considered any parents who may wish to access reproductive services. They advised that IVF should be made available on the National Health Service and called for legislation that legally defines which party is the mother when a child is born with the help of egg donation.[8]

Outside of her work in science, McLaren had an interest in politics. Although she had a privileged background, being the daughter of a baron, she was an active socialist and a member of the Communist Party. At a time when the Cold War played a large part in current affairs, McLaren supported her fellow scientists in the Soviet Union.[9] She was also a member of the Pugwash Conferences that work towards the social responsibility that needs to go hand in hand with science. As a mother of three, McLaren was also passionate about women's rights and access to childcare for working mothers. Her job as a research scientist allowed flexible working hours which could accommodate family life; however, she was empathetic enough to appreciate that not all fields came with this privilege.

Viewing her life-long interest in genetics through a much-needed perspective of conservation and sustainability, McLaren became one of the co-founders of the Frozen Ark project in 2004. This is an organisation, still in operation, that safeguards DNA of endangered species.

On 7 July 2007, McLaren and Michie were killed in a car accident; a sad and sudden end for a brilliant mother, scientist, and colleague. However, the months leading up to her death were full of celebration and recognition of a career that had spanned nearly six decades. Friends and colleagues from near and far gathered in Cambridge to observe her eightieth birthday with a symposium on germ cells and stem cells. The following month, McLaren was a joint recipient of the March of Dimes Prize in Developmental Biology. This prestigious award was given in acknowledgement of the work that McLaren and Janet Rossant had done in using mouse models to understand mammalian reproduction and development. So, although her plans to travel the world for both professional and personal reasons were cut short, this illustrates how the life of Anne McLaren was a tremendously inspiring one. The impact she had on policy makers as well as her peers and budding scientists meant that her sphere of influence was wide indeed. She was conscious of the implications of new technology and led the way in choosing the path of social responsibility – something that will no doubt become ever more important in our constantly developing world.

Laura Bassi (1711–1778)
Physics
Italy

Although much of the work of eighteenth-century polymath Laura Bassi has been lost to history, there can be no doubt about the influence she had on changing views on female competency. Several men who went on to make great scientific advancements sent work to Bassi for her approval, such was her standing. Making a name for herself as the woman who understood Newton's work on mechanics, Bassi's story is an unusually positive example of female achievement being not only encouraged but commended to the point that she became the first female to hold a teaching position at a European university.[1]

Bassi did not come from a particularly wealthy family but a few resources and a helping hand set her up for success. She was lucky enough to start her education aged 5 and clearly had a spark, as Gaetano Tacconi, the family doctor, was so impressed by her potential that he offered to tutor her himself. By her late teens, Bassi was showing great aptitude for natural philosophy, metaphysics, Latin, French, and logic. Although her interest in Newtonian physics caused her and her tutor to drift apart, Bassi attracted the attention of a prestigious member of society who agreed to act as her official sponsor. Prospero Lambertini would later become the Archbishop of Bologna, and Pope Benedict XIV after that. His standing enabled him to arrange a series of public debates in which Bassi could take part. Throughout these events in 1732, Bassi impressed her audience (who by this time were very aware of this intellectual wonder woman), and she managed to hold her own and defend forty-nine theses. Because of her performance, she was then awarded honorary membership of the Bologna Academy of Sciences, shortly followed by a PhD. This made her the first recorded woman to hold a doctorate in a scientific subject. The next set of theses she defended were about the properties of water; this was the event that started her career in science. In what was a shocking turn of events given the social context of the time, Bassi was given a professorship.

Young and eager to start her new life as an academic and teacher, Bassi was unaware that she was about to come across a significant barrier. Although this position had been voluntarily offered to Bassi by the University of Bologna,[2] the lecturing hours allocated to her were far fewer than her male peers. Despite challenging this, her request to increase her teaching time was not acknowledged. Not content to sit on her laurels, Bassi then went above and beyond, embarking on an ambitious goal to fill any gaps in her knowledge, primarily mathematics. She enlisted the help of the best scholars around, learning everything she could to strengthen her position. However, in 1738 she married Giovanni Giuseppe Verratti. They had five surviving children,[3] but the marriage also provided an alternative option for her teaching career. As she had male students, teaching from home as an unmarried woman may have caused social issues regarding her reputation, so having a husband afforded her some protection. She would continue to teach experimental physics in this way for nearly thirty years,[4]

Bassi's career spanned an exciting time in physics. Given that Newton had published his ideas on optics and his ground-breaking work 'Principia' in 1672 and 1687 respectively, Newtonian physics was not a mainstream area of study in Italy around the time of Bassi's teaching debut. Newton's publications offered a deep insight into how the world works, using mathematics to combine the theories of scientists such as Galileo and Descartes, who had come before him. He showed how light can be split into different colours when passed through a prism, how mirrors can be used to build telescopes, and how orbiting planets move. On a more day-to-day level, Newton explained the core principles of how objects move, both through free space and restrictive mediums such as fluids. Though it is easy to see how these principles fascinated scholars of the day, Bassi played an important role in presenting the information to her students in an accessible way. She became one of the first academics in Italy to teach natural philosophy based on Newton's work; by the time she died, it was a fundamental area of physics and considered the launch pad for further exploration of many of the natural sciences.

In the 1760s, Bassi started a collaboration with her husband, who was also a professor at the university. Together, they made Bologna a city renowned for experimental research in electricity. Alessandro Volta, who would go on to invent the battery, placed great value on her as a mentor, even sending her his earliest publications for review. Prestige followed Bassi until the end of her days. Just a couple of years before her death, she was appointed as professor of experimental physics at the Bologna Institute. Her husband was her assistant;[5] as power dynamics go, this was unusual and even more so given the era they lived in.

Anyone who is lucky enough to have experienced the mentorship of an inspiring teacher will know what a difference their influence can make, and how they have the potential to change the course of your life. It is difficult to estimate how many women Laura Bassi had an effect on whether directly or indirectly. How many fathers she convinced to let their daughters have an education, against the cultural norms of the era. How many male leaders she encouraged to take the professional ambition of women seriously. The influence Laura Bassi's name had on generations to come should be remembered and celebrated.

Further Reading

Caroline Herschel – a German astronomer who is believed to be the first salaried female scientist.

Alyce Gullattee (1928–2020)
Psychiatry
United States

Alyce Gullattee (née Chenault) was born on 28 June 1928 in Detroit, Michigan. She grew up with over ten siblings; frequent illness in the family was the reason that from a young age she aspired to go to medical school and become a doctor.[1] As well as having a passion for helping children, Gullattee was also interested in conditions common in elderly populations, which led her to specialise in psychiatry.

Gullattee's first degree was in zoology. She studied at the University of California, Santa Barbara in 1956,[2] before earning a medical degree from Howard University, Washington DC in 1964.[3] From her childhood, Gullattee had witnessed the racial injustice common in her own and surrounding black communities. Having joined the National Association for the Advancement of Colored People during high school, she had grown up with an understanding of how to strive for change and maximise the chance of success. Gullattee had come to the realisation that words and actions alone were sometimes not enough to educate people. She went on to use her position and expertise in the medical field to build a targeted, methodical approach. To Gullattee's disappointment, prejudice was still rife in the medical profession; black and minority medical students were regarded differently to white medical students which often affected the involvement they were able to have. In 1964, she helped found the Student National Medical Association, which to this day provides practical support to medical students by offering opportunities to share work and encourage the development of culturally sensitive leaders. It was suggested that this new organisation for minority groups should pay membership fees and become part of the existing association for medical students. However, Gullattee opposed this bitterly. Her peers who were equally as qualified as their white counterparts were frequently discriminated against, so there was a strong desire to set up a separate organisation.[4] Eventually, it was decided that they would become a subsection of the National Medical Association which represented African American physicians. This differentiation was vital because Gullattee's vision was to support not only medical students but also underserved communities that did not seem to be a priority for existing professional organisations.

Four years after completing her medical degree, she joined the faculty of Howard University and started work in the Department of Neuropsychiatry. Gullattee traced her first experience of witnessing substance abuse back to her final year as an undergraduate. Seeing a group of primary school children growing cannabis and distributing it in cigarette form to their fellow students opened her eyes to the importance of teaching the negative effects of drug use.[5] As her responsibilities started to involve attending court hearings, it became apparent that there was a strong link between juvenile crime and substance abuse. Gullattee's work was behind much of the education that became available for those working with children around the Washington DC area, and her audience ranged from school staff to juvenile court judges. Due to her expertise, she became director of the Institute on Drug Abuse and Addiction and worked with patients affected by the 'crack epidemic'.

As seen throughout her impressive career that spanned five decades, Gullattee had the patience and empathy to understand the people whom wider society might dismiss as being 'difficult'. The renowned psychiatrist cared a great deal for her patients; perhaps for some of them she was the only person who would listen and give them the time of day. Not content with lumping people into categories, she took the time and got to know the patients she worked with. When the 1975 Washington DC Human Kindness Day festival turned violent after groups of young people initiated robberies and assault, Gullattee explained that there was an unfortunate inevitability to the events that transpired that evening. The city was divided and racial tensions were running high; she urged that rather than people being at fault, the system that disproportionately raised up white policy makers to make decisions was isolating young people from minority backgrounds. These children were growing up in a hostile environment which led to a lot of anger.[6] Further proof of her compassion and kindness to those around her was when a young woman addicted to heroin gave birth to her baby in the hospital where Gullattee was on duty. A blanket was knitted for the newborn by the doctor and she agreed to be listed as her next of kin.[7]

On a national level Gullattee's medical prowess was appreciated by Presidents Nixon, Ford, and Carter; serving on several White House committees.[8] Her civil rights activism and knowledge of psychiatry undoubtedly touch the lives of countless people, and pushed for fairer treatment of the most vulnerable groups in Washington DC.

Humanity lost a great pioneer when Gullattee passed away from COVID-19 in 2020, but she will continue to be a vital role model to those aspiring to emulate the level of kindness she so selflessly shared with her community, state, and country.

Chapter 4

Global Health

Further Reading

James Barry – Although this military surgeon who lived between 1789–1865 spent his adult life as a man and is therefore not within the scope of this book, his story of social mobility, humanity, and medical reform deserves a special mention. As well as being a healthcare icon and a vital role model for the modern-day trans community, Barry is equally remembered for being a doctor who fought for better care for soldiers and civilians throughout the British colonies regardless of their social status or ethnicity. This physician whom some sources state as conducting one of the first successful Caesarean sections in Africa[1], was also notoriously scrupulous about sanitation and hygiene. He had no tolerance for people who cared more about how medicine could make them profit than the effect it had on patients and took a stance against the sale of medicines by unqualified vendors. It was clear that Barry had a genuine passion for public health; his concern for the welfare of the mentally ill, enslaved people, and prisoners[2] was ahead of his time and not appreciated by many of those around him.

Handwriting analysis and a letter signed by Barry himself provides evidence that he had previously been known as Margaret Bulkley[3] and that his gender assigned at birth had been female. This change in identity occurred when Barry registered as a medical student at the University of Edinburgh after several failed attempts to find work as a governess. Whether or not his lifestyle change was due to reasons other than accessing higher education, Barry's long career that culminated in his prestigious appointment as inspector general of hospitals was hugely significant in proving that a person who was born female was capable of being a successful medic. A new age was dawning for British medicine. The same year James Barry died, pioneer Elizabeth Garrett finally qualified as Britain's first female physician after a huge struggle that involved heated debates and the threat of lawsuits. The cracks were beginning to show in the predominantly male world of British medicine and with the support of some important global allies, it would eventually crumble.

Patricia Bath (1942–2019)
Ophthalmology
United States

Patricia Era Bath is primarily celebrated for being an ophthalmologist who pioneered laser cataract surgery. Her career was full of firsts and included memberships to institutions, job roles, and applying for patents. On top of all this, she found time to inspire countless people with her important humanitarian work; it is no wonder that according to her obituary in *The Lancet*, her friends and colleagues described her as 'a role model for future generations'.[1]

Bath was born in 1942 in Harlem, New York and grew up with a multicultural background; her mother was a descendant of both slaves and Native American Cherokee and her father was an immigrant from Trinidad and Tobago. Both parents were supportive; Bath was gifted a chemistry set when she started to show an interest in science, and her mother would later take on work as a housekeeper to help fund the education of her children. Bath also found support at school; when her teachers noticed her natural talent in maths and science they encouraged her to take extra biology courses to hone her research skills. This, combined with a dose of inspiration from Albert Schweitzer, a doctor who set up a hospital and offered medical aid to thousands of local patients in Lambaréné in modern-day Gabon, prompted Bath to apply for a summer school scholarship funded by the National Science Foundation. Her application was accepted in 1959 and she was given the opportunity to carry out research that helped conclude that cancer is an illness that causes loss of weight through degradation of fat and muscle mass and that the growth of tumours is a symptom of the disease.[2] At this point, she was just 16 years old. Before she even completed high school, Bath's work (which had contributed to these findings) had been included in a scientific paper and she was also one of ten people in the US to win the Mademoiselle Magazine Merit Award.[3] Her education continued; she studied physics and chemistry at Hunter College in New York before moving on to the Howard University School of Medicine.

Bath spoke highly of her time at Howard University, a historically black institution; as it was the first time she was exposed to people of colour who led the way in their academic fields.[4] In a conversation recorded for The Foundation of the American Academy of Ophthalmology, Bath recalls that she was one of eight or ten students inspired to specialise in ophthalmology by Dr Lois Young, who was acting Chief of Ophthalmology at the time. Young had risen to her position against all odds, having been initially denied training at her university of choice because of racial discrimination. It is easy to understand how her influence had such a positive impact on the confidence and motivation of students that it would eventually shape their future careers. Bath graduated with honours in 1968, a year not only significant on a personal level but for the whole of the US and, by extension, the world. Reverend Martin Luther King Jr was assassinated on 4 April. This triggered turbulent social unrest whilst supporters of the civil rights movement attempted to reorganise themselves after the tragic death of their leader and find a way to carry out his legacy. The Poor People's Campaign was one plan the reverend had in store; at its heart, it was a call for justice, both in terms of economic and human rights for poor Americans from a range of backgrounds. Data from the US Census Bureau shows that in 1959, 22 per cent of the population lived below the poverty line.[5] Bath played her part by leading a team of students from her university to provide healthcare services for the campaign.

There is no doubt that her involvement with the Poor People's Campaign opened Bath's eyes to the low level of healthcare available to the poorest in society, but when she returned to New York to complete her training in ophthalmology she started to develop another uncomfortable realisation. Having worked at both Harlem Hospital and the eye clinic at Columbia University, she started to notice that there were many more patients who suffered from severe visual impairment in Harlem.[6] Combining her humanitarian soul with her scientific mind, Bath decided to use state blindness registries to conduct an unfunded, retrospective study to see whether there was any reason for this difference, based purely on the location. She started to collect data from both clinics and eventually came to the distressing conclusion that in the US, blindness was twice as common in the black population as it was in the white population.[7]

The reasons behind this are complex and perhaps not fully understood to this day; however, Bath's experiences of growing up in Harlem and studying abroad during her degree had provided her with some insight into methods that could offer effective and, more importantly, quick improvements. In the late 1970s, Bath published her findings and proposed solutions in her paper, 'Rationale for a Programme in Community Ophthalmology'. Her research suggested that, in almost every state, black and minority races had a much lower level of healthcare access than the white population even though they were more often affected by severe eye diseases.[8] The data did not just include adults; it also showed that black children were less likely to be issued with glasses than white children. The findings presented in the study partly explained the statistics related to blindness and also glaucoma, a condition in which pressure causes damage to the optic nerve. Bath suggested an approach that combined ophthalmology with public health, community medicine, health education, epidemiology, and biostatistics, which she presented to the American Public Health Association. She ensured that a greater level of healthcare access was made available to the people of Harlem by arranging for professors at Columbia University to operate on blind patients free of charge. Bath herself volunteered her services as assistant surgeon. However, as dedicated as she was to the patients of New York, her vision was further reaching; as she mentions in the abstract of her paper on community ophthalmology, her rationale is discussed on not just a national level but also international. Today, community ophthalmology is used all over the world; by screening and testing the vision of patients within their communities, thousands of people are saved from going blind.

Between completing her internship and publishing this study, Bath became the first black woman to complete an ophthalmology residency at New York University. During this time, she also got married and had a daughter. Corneal transplants and keratoprosthesis surgery – the replacement of a diseased cornea with an artificial cornea – eventually became her focus. Two years later, Bath became a pioneer yet again when she accepted two jobs; one as an assistant professor of surgery at Charles R. Drew University and another as assistant professor of ophthalmology at the University of California, Los Angeles (UCLA). The latter meant that she was the first female faculty member at the Jules Stein Eye Institute; this seemed to cause a few logistical challenges for her employer. In various interviews, Bath recalled that rather than arranging for her to share an office with her male counterpart as was the norm, she was given an office in the basement of the building next to the laboratory animals.[9] They eventually reached a compromise and Bath agreed to share an office with the secretaries.[10] As she said herself, it would have been perfectly understandable to refuse or raise a complaint, but Bath decided to overlook the issue with initial office arrangements and focus on the work that needed to be done. Very quickly, she co-founded the King-Drew-UCLA Ophthalmology Residency Program at Martin Luther King Jr Hospital. Bath would go on to become the chair and the first woman in the US to lead a residency programme in her field,[11] and those who graduated from the programme became the professionals who would go on to provide care and screening as well as practical support for blindness prevention. Bath also built up a legacy at the Jules Stein Institute by founding the Keratoprosthesis Program, which runs to this day, using surgical techniques to restore eyesight to those suffering from eye diseases.

There is no doubt that being employed by two academic institutions put considerable constraints on Bath and perhaps did not offer the opportunity for creativity and independent thinking that she was clearly so capable of. In 1976, she founded the American Institute for the Prevention of Blindness (AiPB) with two colleagues, doctors J. Alfred Cannon and Aaron Ifekwunigwe. Built on the beliefs of its founders, this non-profit organisation operates under the motto that eyesight is a basic human right and has been responsible for many sight-related outreach programmes in underserved communities all over the world.[12] By 1986, Bath had offered twelve years of service to Drew University and this entitled her to a sabbatical. She chose to work in Europe, visiting the Rothschild Eye Institute in Paris, the Institute of Technology in Loughborough and the Laser Medical Centre in Berlin.

Five years previously, Bath had devised a novel piece of equipment and method that could quickly remove cataracts from the eyes of patients. The traditional process for achieving this was much riskier than it is today so Bath's laserphaco probe, which used a laser to vaporise and remove the cataract in a minimally invasive way, was truly ground breaking. Despite having had this idea for years, she had not had access to the appropriate lasers and testing equipment; now using state of the art laboratories in

European clinics, Bath was able conduct early studies into laser cataract surgery to prove the credibility of her invention. After years of hard work, the device was patented in May 1988.[13] This brought with it yet another first for this ambitious lady: she became the first black woman to be awarded a patent for medical purposes.[14] She would go on to receive four more patents, even filing for them herself when she became frustrated with the slow progress of her patent attorneys.[15] Bath's achievements with the AiPB and cataract surgery are two chapters in her life that showcase her as not just an excellent researcher but also an innovator.

Bath retired from UCLA in 1993 and by the end of her career had written over 100 papers.[16] Her life was littered with trials and people who didn't believe in her ability to change the world; from the director who pronounced her invention of the laserphaco probe as 'impossible',[17] to certain delegates of the 1976 American Public Health Association who didn't see the relevance of an ophthalmologist being present when she presented her study on community ophthalmology.[18] It seems fitting to end with a quote from Bath herself. 'Believe in the power of truth ... do not allow your mind to be imprisoned by majority thinking.' In our current world of turbulent politics, internet influencers and fake news, perhaps her words are more relevant than ever.

Kin Yamei (1864–1934)
Medicine
China, United States

Kin Yamei was a doctor who used her multicultural upbringing to change perceptions and encourage the sharing of medical knowledge and resources. Born in Chekiang Province, China, in 1864, Kin started life surrounded by religion. Her parents were independent thinkers, having counter-culturally converted to Christianity and chosen a love match over the more traditional route of an arranged marriage.[1] Tragically, when she was only a couple of years old, an epidemic swept through Kin's hometown claiming the lives of both her parents by the August of 1866.[2] With no living family or friends to take care of Kin and her brother, they were adopted by American missionaries, Dr Divie McCartee and his wife Juana, who had been acquainted with the children's late parents.

In the early 1870s, Kin's foster parents took up residence in Japan for several years and both played a role in providing her with a broad education. McCartee worked for the Ministry of Education and lectured on a range of scientific subjects such as anatomy and botany, whilst his wife had a good knowledge of English literature and language; between them they were able to tailor their knowledge to pass on to their daughter. Although Kin had expressed the wish to become a teacher,[3] it was eventually natural sciences that captured her interest and directed her path to becoming a doctor who would cross cultural borders.

The McCartees were open-minded people who had dedicated decades to learning about East Asian culture and medicine. However, their home country was not so friendly. There was hostility towards Chinese immigrant workers and a widespread belief that prostitution and the links to (recently prohibited) slavery that came with it was an issue more common in immigrant communities. This discontent had led to the passing of the Page Act in 1875, a restrictive immigration law that prohibited Chinese women from entering the country. As well as social challenges, it was founded on an inherent lack of understanding of Chinese culture plus some outlandish beliefs popularised by the American Medical Association. As Chinese women were thought to carry germs that could cause death in white people, the support for this Act became fuelled by ignorant health concerns as well as objections to morality. The implementation of this law was arguably worse, with all female Chinese immigrants being asked a series of humiliating questions to 'examine' their moral values and placed under scrutiny by immigration officers instructed to establish whether each woman was a prostitute. Attitudes towards Chinese people did not improve; in 1882, the Chinese Exclusion Ban would extend the immigration restrictions to men as well as women. Only those with desirable skills were provided legal access to the US.

It was in May 1880 when the McCartees moved to the US and Kin entered a world where she might be unfavourably judged because of her ethnicity and gender. She completed a year at the Rye Seminary then embarked on her career path around 1882, applying to the Woman's Medical College of New York Infirmary. Kin graduated three years later, becoming the first Chinese woman to earn a medical degree in the US,[4] and continued her studies in New York, Philadelphia, and Washington. As well as studying, she completed two residencies, became skilled at photographing samples under a microscope, and published an article in the *New York Medical Journal*.

With her education complete, Kin felt drawn to return to China. Perhaps following in the footsteps of both her biological and adopted fathers, her chosen role was as a medical missionary. However, her placement did not last long. Firstly, it appears she did not receive a warm welcome; due to the mission medical adviser's dislike of female doctors, Kin was not offered accommodation[5] or assistance at work. As if enduring this prejudice wasn't tough enough, she subsequently caught malaria. In search of a more calming atmosphere in which to recover, Kin left for Japan not much longer than a year after she had arrived in China. After regaining her strength, she is thought to have worked with another missionary group in the city of Kobe. With less of a critical eye upon her, Kin seemed to thrive in this environment and provided a great service to local people by setting up a clinic for women and children. Within four months of opening, she had over sixty patients[6] and was so under-resourced that she had to cover multiple roles whenever disease rates increased. As well as treating and nursing patients, she educated local doctors and midwives by passing on techniques and knowledge she had learned during her own medical training in the US. After her successful years in Japan, Kin married Hong Kong-born interpreter and merchant Hippolytus Laesola Amador Eça da Silva. Shortly afterwards, the newlyweds moved to Hawaii where their only son, Alexander, was born a year later.

Around the turn of the twentieth century Kin took the role of educator even further, this time giving lectures on Chinese culture. One of her first lectures was given whilst living in Hawaii; from here she went from strength to strength, delighting her listeners with information about the society, people, and medicine of her homeland. She would go on to give talks at the 1904 International Peace Conference, to the Ethical Culture Society, and the socialists of the Cooper Union.[7] Soon she was in popular demand and travelled all over the US for speaking appointments. Keeping busy may well have been a coping mechanism for dealing with her personal life, which was turbulent around this time. Kin was divorced from her husband in 1904. This was initiated by Eça da Silva and it transpired that he did so with the claim that his wife had abandoned him. Kin's rather simple explanation was that her husband was not 'up to date'.[8] Whether or not this guarded statement was linked to the events that occurred later that year is unclear; Eça da Silva engagement to two different young ladies came to light and he was also arrested (though later acquitted) after he was accused of importing women into the country for nefarious purposes.[9] It is telling that when travelling further afield in previous years, Kin left their son in the care of friends rather than his father.

Whilst Eça da Silva was in the media for all the wrong reasons, his former wife enjoyed good publicity and her skills as a speaker were even praised in an article published in the *New York Times*.

With Alexander enrolled at boarding school, Kin started looking for her next career opportunity and this time found it in her home country. She returned to China in 1905, initially settling in Sichuan Province, where her experience was very different from the last time she had practised medicine in the country. Kin had a pivotal role in opening several clinics, which allowed locals access to a level of healthcare they had not previously enjoyed. A year into her work, the central government started to promote female education, which fell nicely in line with Kin's values and skill set. In Tianjin, she oversaw the Government Women's Hospital, then founded and organised the building of the Northern Medical School for Women, where she taught modern nursing techniques. The school had a major focus on enrolling girls from the region of Zhili, especially those who came from underprivileged backgrounds. Because of this, Kin's work went beyond simply providing these children with an education; it was an opportunity to learn a vocation which would increase social mobility. Their lessons on obstetrics, public health, and pharmacology would improve the lives and livelihoods of people throughout the country. From an article in the *Daily Tribune*, dated 1909, we know that tuition fees were paid for by the government and that students would have a salaried position after graduating.[10] Providing and managing the education of these girls who would go on to play a part in nationwide healthcare development would be just one of Kin's many legacies.

When Kin returned to the US in 1911, she took with her a recent graduate from the nursing school. Her former student planned to study medicine at Johns Hopkins University and Kin used this visit to learn about the latest techniques and technological developments in hospitals. By the time the rest of the world was edging ever closer to the start of the First World War, China had already been involved in a number of historically significant events and was experiencing the end of its last imperial dynasty. Whilst the tail end of the Qing dynasty was mired with rebellion and conflict, Kin did benefit from the education reforms that were implemented in the name of improving prospects for the Chinese population. A Chicago-based magazine published an article stating that Kin had been responsible for promoting the emigration of Chinese women to the US to further their medical studies. By all accounts, she had encouraged as many as forty women to do this.[11] Shortly before the beginning of the First World War, Kin started to have dealings with the US Department of Agriculture (USDA). She introduced the Bureau of Plant Industry to a number of plants native to China. Her work with USDA continued throughout the war, culminating in her employment as a government agent. Her brief was to find Chinese crops that could act as alternatives to red meat, wheat, and vegetable oils; foods that were likely to be affected by disruption to the supply chain during the war. Kin discovered that soy bean derivatives could meet many of the criteria. And there were other reasons for promoting the use of soy in the American diet at the time. A diet high in meat and dairy is affected by the inherent energy losses of a longer food chain. Kin showed an understanding that the fewer steps there are in the food chain, the more efficient the process is. The interest and engagement in this idea was limited at the time, but it is recognised to have become more important as the modern world, with its limited resources, is grappling with the challenge of trying to feed an increasing global population.[12] It is interesting to note that up to this point in her career, Kin had been using Western knowledge and techniques to share with her Chinese students and colleagues, but her role was reversed during the war. In America's time of need, they turned to China's wisdom and its perspective on agriculture that had been honed over the centuries. As her son was fighting on the front line in Europe, working with the USDA was Kin's way of contributing to the war effort. Unfortunately, Alexander's service to his country ended abruptly. He was killed at the Second Battle of the Somme, mere months before the Armistice was signed.[13]

With both her foster parents also deceased, Kin moved back to China to live a life of relative privacy. She still contributed to community projects such as the Peiping Municipal Orphanage and the Chingho Village Experimental Centre,[14] and also led an active social life. Her last home was in Peking where she would eventually die in 1934 from pneumonia.[15]

By the end of Kin's life, her experiences and knowledge earned her the reputation of a societal matriarch and her incredible stories would live on in the writings of her lodger Jaroslav Prusek.

Kamala Sohonie (1912–1998)
Biochemistry
India

The beginnings of British Rule in the country we now know as India began as early as 1612. A treaty between between the Mughal Emperor and the East India Company, under the guise of advancing trade, would later force dominion over parts of the subcontinent which ultimately led to the British Raj. As one can imagine, being ruled by a country halfway across the world that is so culturally different could often lead to social strain and tensions. Hierarchy, violence, and power were common, with many Europeans talking down to the Indian population, insisting they were unable to match the intellect of people from the West.[1]

From the 1890s, decades into British crown rule, Jagadish Chandra Bose and Prafulla Chandra Ray worked hard to show that this racist viewpoint was incorrect.[2]

In 1921, Mahatma Gandhi became leader of the Indian National Congress.

And in 1933, Kamala Sohonie (née Bhagvat) would start a revolution that was on a smaller scale but every bit as important.

Not much has been documented about Sohonie's early life, but we know she was born in 1912 in Indore, Madhya Pradesh. Having a father and uncle who were chemistry alumni of the Tata Institute of Sciences, it can be assumed that education was regarded as important. Sohonie was accepted to Bombay University, where she proceeded to pass the exams for her bachelor's degree in chemistry and physics and come top of her class.[3] Having achieved this, her plan was to complete a master's degree at the Indian Institute of Science (IISc). The institute was directed by Nobel laureate Chandrasekhara Venkata Raman, the man who had discovered that the wavelength and amplitude of light changes when entering a different medium, therefore altering its colour. By examining the change that occurs when light passes from air to water, he explained why the sea is blue.[4] It must have been with some apprehension that Sohonie handed in her application, and perhaps

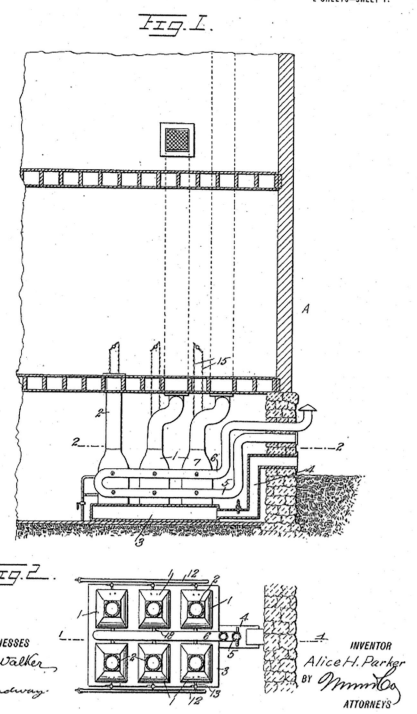

Alice Parker furnace. (*Courtesy US Patents and Trademark Office (US1325905A) - Public Domain*)

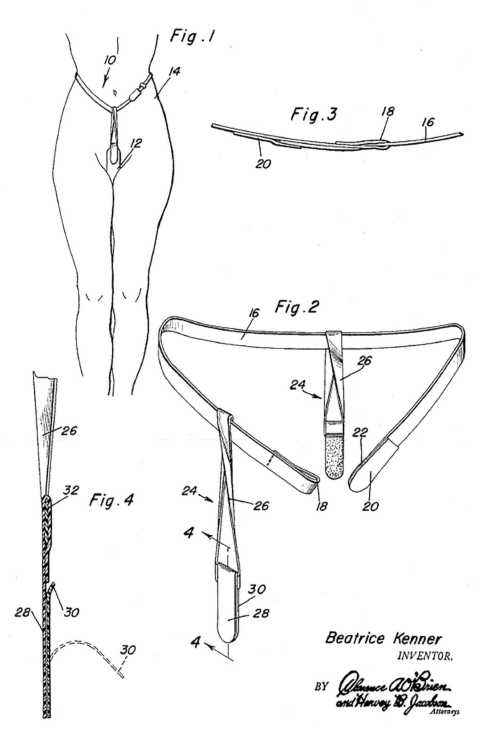

Beatrice Kenner sanitary belt. (*Courtesy US Patents and Trademark Office (US2745406A) - Public Domain*)

Alyce Gullattee.
(*Courtesy Brigette Waters-Turner*)

Japanese Internment Poster. (*Photograph No. 210-G-A39 "San Francisco, California. Exclusion Order posted at First and Front Streets directing removal of persons of Japanese ancestry from the first San Francisco section to be effected by the evacuation." November 1942; Records of the War Relocation Authority, 1941 - 1989; National Archives at College Park - Still Pictures (RDSS)*)

Caroline Haslett. (*Courtesy Institute of Engineering and Technology*)

Kamala Sohonie. (*Courtesy Indian Academy of Sciences*)

Inge Lehmann. (*Courtesy Royal Danish Library*)

Kin Yamei. (*Courtesy James Kay McGregor,* "*Yamei Kin and her Mission to the Chinese People*" *The Craftsman* (*November 1905*))

Lise Meitner. (*Unknown Copyright - First Printed 1899*)

Lillian Moller. (*Gilbreth Courtesy Special Collections and University Archives-Rutgers University Libraries*)

Marion McQuillan. (*Courtesy Institute of Materials, Mining and Minerals*)

Mary Anderson. (*Unknown Copyright - No Known Photographer*)

Patricia Bath. (*Courtesy of the U.S. National Library of Medicine. https://www.nlm.nih.gov/copyright.html*)

Rachel Carson. (*Credit: USFWS*)

Stephanie Kwolek. (*Courtesy Science History Institute; Photographer Harry Kalish*)

Virginia Apgar. (*Courtesy Library of Congress, Prints and Photographs Division, NYWT&S Collection, LC-USZ62-131534*)

rightly so because it was promptly rejected by Raman on account of her being female.[5] Sohonie rebuffed his rejection. Although she has been described as a 'quiet, unassuming person',[6] this woman was not afraid to stand up against injustice, as she would go on to prove during the freedom movement. Rather than kicking and shouting, Sohonie took inspiration from Mahatma Gandhi and practised non-violent protest, *Satyagraha*, outside Raman's office. Unable to provide a logical answer for his reasoning, Raman gave in and the gender barrier for all future female applicants was removed over time. Despite this important victory, Sohonie was still subjected to prejudice, as the following restrictions were listed as a condition of her admittance:[7]

1. Sohonie would be on probation.
2. Unlike the male students, her future at the institute would depend on Raman's satisfaction with her work after one year.
3. She must not spoil the environment, which has been interpreted by some as 'not distracting her male colleagues'.

These terms clearly stayed with Sohonie for the rest of her life, as she reminisced about this period at an event organised by the Indian Women Scientists' Association (IWSA) in 1997. Making it clear that whilst respecting Raman as a scientist, she found him to be narrow-minded. [8]

Fortunately, Sohonie was able to prove to her boss and male peers that women can be good researchers. This was just as well because she had already announced to Raman that she would achieve a distinction! For the experiment detailed in her first paper, Sohonie chemically processed nine indigenous pulses. It was already known that pulses contained protein, a food group that is vital for the building and healing of cells. However, there is also a fraction that is made up of 'non-protein nitrogen'. Nitrogen is a key component of amino acids, commonly known as the building blocks of proteins. Until this point, no one had researched the importance of this part of pulses, so that is exactly what Sohonie did. She paid special attention to heating the solution; as she states in her paper, this is the closest representation of it being cooked, which is how children and the elderly in particular consume this food.[9] In summary, the way in which pulses are prepared can alter the availability of nitrogen and by extension amino acids – and therefore protein. By cooking the pulses, the body can spend less energy digesting the protein and absorbing the nutrients. Sohonie then went on to look at the digestibility of milks. In fact, by the time she handed in her dissertation in 1936, two of the five research articles she had written were of a high enough standard to be published in *Biochemical Journal*.[10] Needless to say, she achieved her distinction.

Sohonie's dedication to science earned her a scholarship to Cambridge University in 1937, where she studied with Dr Derek Richter, then Dr Robin Hill who was known for his work on photosynthesis. Hill and Sohonie focused their research on a water-soluble protein called Cytochrome c. It is found in the mitochondria which are commonly referred to as the 'powerhouses' of the cell. Cytochrome c plays a vital role in the planned death of animal cells, which is important as continued and uncontrolled growth is harmful, but their research showed that this protein was also present in plant cells.[11] It is reported that Sohonie handed in her remarkably short thesis of forty pages[12] less than two years after starting work in Cambridge, making her the first Indian woman to receive a PhD in a scientific discipline. During this time, she had also managed to secure two scholarships: one involved working on oxidation with Nobel laureate Professor Frederik Hopkins, who had written about the dietary importance of vitamins; the other was a travelling fellowship from the American Federation of University Women. There is very little documentation about her life during these years, but it was surely an encouraging experience to meet other female scientists from a range of countries. Although she had acquired new contacts and interested a number of foreign companies enough to receive profitable job offers, the now very qualified and experienced Sohonie moved back to India in 1939. Her son later commented on his mother's patriotism, suggesting that she returned to show support for the freedom movement that was rapidly gathering pace. Sohonie may well have become an international household name had she accepted a job at a large pharmaceutical company – her son went as far to say that she may have won a Nobel Prize – but the work she became known for provided some desperately needed aid to the poorest people in her country.

We must remember that she was returning to a pre-independent India and, whilst it is easy to observe her choices in hindsight, it is unlikely that we will ever understand the gravity of what she felt she was giving up – or perhaps gaining – from going home. What an example she set, choosing humanity and loyalty over money and international prestige.

Lady Hardinge College in New Delhi had recently opened a new department of biochemistry and Sohonie took a position as its head of department when she arrived back in India. It wasn't long before she was moving up the career ladder, researching vitamins as the assistant director at the Nutrition Research Laboratory in Coonoor.

For both Sohonie and her homeland, 1947 was a life-changing year. As she was joined in matrimony to M. V. Sohonie and moved to Bombay to start a life with him, the Partition of India divided the country into two independent states. The countries of India and Pakistan at midnight on 15 August and put an end to the British Raj. Whilst the details are beyond the scope of this book, with widespread violence and a death toll estimated between 200,000 and 2 million,[13] it was a turbulent time that would have undoubtedly had a deep effect on Sohonie. From an article written by Anirban Mitra, we know that she gave up wearing jewellery and donned rough Khadi sarees, which was part of Gandhi's widespread protest against the British Raj trying to sell expensive cloth. In Bombay, she became a professor of biochemistry at the Royal Institute of Science and went back to her roots of studying legumes and other foods commonly eaten by poorer communities. We now live in a world where many people are health-conscious; calories are counted, cholesterol is reduced, and certain food groups are endorsed or avoided. But for Sohonie and the wider scientific community, this was pioneering research. She had the knowledge and the authority to use this information for good, and to help her impoverished countrymen who struggled to stay healthy because large areas of nutrition were not properly understood.

Whilst historically many women were required to choose between the distinct paths of career or family, Sohonie, in open defiance of societal prejudice and expectation, gave birth in 1950 and again in 1951 then carried on with her work at the Royal Institute of Science. For this trailblazing biochemist was not done yet. She eventually became director of the institute and went on to research the nutritional value of a widely consumed drink called Neera, at the request of the first president of India, Rajendra Prasad. Her group discovered that this drink, made from the sap of flowers from various species of the Toddy Palm, contains a beneficial amount of Vitamin A, Vitamin C, and iron. Some food loses nutritional value when cooked or processed, but they also managed to find that that the vitamins and minerals in Neera were still present even after condensing the sap into sugar.[14]

Sohonie's influence grew further still as she became a member of the Consumer Guidance Society of India (CGSI) and served as its president.[15] One of only sixteen people to hold this title since the society was founded in 1966, the fact that she was elected suggests a refreshing development since the days when women were excluded from joining certain research facilities. Her humanitarian legacy lives on in the consumer safety articles she wrote for *Keemat* magazine, showing that she had become a trusted and reputable advisor as well as an excellent researcher, scientist and teacher. As Sohonie explained, Indian women were commonly limited to the domestic chores of childcare and housework and had no place outside the home.[16] Kamala Sohonie proved that not only do women have a place in the world, they have the ability to change it.

Further Reading

Elsie Widdowson – a British dietician who is best known for her role in advising the British government on the nutritional supplements required during wartime rationing.

Jane Cooke Wright (1919–2013)
Oncology, Medicine
United States

Jane Cooke Wright came from an impressive family of firsts. Her father, Louis Wright, had managed to get into Harvard Medical School despite strong racial discrimination[1] and after qualifying had been the first African American surgeon at both Harlem Hospital and the New York Police Department. His own father had been born into slavery[2] but graduated with a degree from the first medical school in the South to accept black students,[3] and his stepfather had been the first African American to earn a degree from Yale Medical School.[4] Therefore, Jane Wright had plenty of inspirational role models as she built a successful career in surgery and cancer research; a career which has touched the lives of people all over the world.

Although she initially set her sights on becoming an artist, Wright soon changed tack and followed in the footsteps of her forefathers, earning a scholarship to New York Medical College. Graduating in 1945, she completed her residency at Harlem Hospital,[5] then stayed there to work with her father at the Cancer Research Foundation he had established. At the time, the only common treatment relied on surgery to remove tumours, but the Wrights explored options of chemical treatment.[6] Because of their bold yet careful approach of testing simultaneously on patients and on tissue culture outside of the body, their work in the field of oncology was pioneering. Their methods were unorthodox; placing patients on clinical trials for chemotherapy was considered to be an 'end of life exercise' at the time, so Wright was a trailblazer in helping to remove stigma and encourage the medical community to see the process as an opportunity to improve future treatment.[7] Gradually, her ideas started to be accepted. Her data showed that some patients experienced remission when treated with anti-cancer agents. In 1951, Wright's team became the first to demonstrate use of the drug methotrexate in solid tumours which are the result of uncontrolled cell division and growth. Previously it had only been known to be effective for the treatment of leukaemia, cancer of the bone marrow. By inhibiting the use of folic acid which is essential for the production of amino acids, DNA, RNA, and proteins that are involved in cell division, the drug was found to suppress tumour growth.[8] Thanks to the extensive research conducted by Wright and her colleagues, it became established as a safe treatment for breast cancer amongst other conditions and was listed in the 2019 World Health Organisation list of essential medicines.[9]

The year following this scientific breakthrough was equally life-changing for Wright, but this time it was painful as she lost her father and partner in research. She became director of the Cancer Research Centre[10] and continued the work she had done with her father, tailoring chemotherapy to be more targeted and effective. Over time, her team's research showed that alternative options to invasive surgery were a real possibility, and that by looking at each patient's cancer tissue, the exact mix of drugs needed for the individual patient could be determined quickly.[11] Wright was causing positive ripples and the developments she had been making had not gone unnoticed. She was appointed to President Lyndon Johnson's Commission of Heart Disease, Cancer, and Stroke in 1964.[12] Given that this happened only a year after Martin Luther King Junior's 'I Have a Dream' speech which called for improved civil rights, it was an extraordinary achievement and indicative of the skill and respect Wright held within her field. This was an essential first step in the continuing journey towards equality and recognition based on talent alone. Wright would go on to hold a number of senior positions after this, becoming the only female of the seven founding members of the American Society of Clinical Oncology, serving for several years on the National Cancer Advisory Board, and becoming the first female president of the New York Cancer Society. She also travelled extensively, sharing her knowledge by leading delegations of oncologists and researchers to Asia, Eastern Europe, and Africa.

Retiring in 1987,[13] Wright could rest safely in the knowledge that her life and career had been a beacon of hope, not only to cancer patients of the past, present, and future but also to women aspiring to work in STEM-based research.

Further Reading

Gertrude Elion – A pharmacologist who revolutionised the way in which pharmaceutical drugs were developed, for which she shared the 1988 Nobel Prize in Physiology or Medicine with colleague George Hitchings. They developed a number of new drugs for conditions including leukaemia, gout, and herpes, and also created an immunosuppressive drug that can prevent rejection during organ transplants.

Alice Ball (1892–1916)
Chemistry
United States

In her saddeningly short life, Alice Ball used her scientific knowledge to change the futures of people suffering from leprosy. The technique she discovered allowed an oil, which had historically been known to sometimes ease the symptoms of the disease, to be administered in a way that was so effective that it became the standard treatment for decades after her death. Some historians suggest that despite the relatively short period of time the Ball Method was used, it provided hope that leprosy was a curable disease and that money spent on research was not spent in vain.

Born on 24 July 1892, in Seattle, Washington, Ball showed a talent for science during her school days. After a brief move to Hawaii, her family returned to Seattle and she enrolled at the University of Washington to study pharmaceutical chemistry and pharmacy. By the time she graduated in 1914 she had earned degrees in both subjects.[1]

Our experience of the Coronavirus pandemic helps put into perspective how important the work of Alice Ball was. Though easy to forget in our scientifically developed world, leprosy was a very real problem and a threat to not only health but livelihoods. Hawaii, where Ball chose to undertake her master's degree in chemistry, has historically been hit hard by various diseases introduced to the islands by sailors, traders, and visitors. The native islanders had no immunity to these diseases that included smallpox, whooping cough, and leprosy; as a result, infection and death rates were high. The government, under pressure from farmers and sugar planters to reduce the risk of their workforce being infected, created two leper colonies that operated on the remote island of Molokai. These colonies were inhabited between 1866 and 1969, so were still extremely significant during the period Ball spent in Hawaii. In his book, *The Colony,* John Tayman explains the fate of leprosy sufferers: how people were arrested and examined if they were thought to have the disease and how they were forced into ships by armed guards and taken to Molokai, as good as dead in the eyes of the law. They would never see their loved ones again. This was a grim sentence for the alleged 8,000-plus men, women, and children who were put in isolation. Ball's discovery had even further-reaching effects than her local island as this brutal method of disease control was by no means unique to Hawaii.

For her master's degree, Ball studied the chemical properties of the kava plant, specifically experimenting with methods of extracting the active ingredient. Her work must have been good as, upon completing her degree, Hawaii College (now the University of Hawaii) offered her a research and teaching position which made her the first female chemistry teacher at the institution. And her employer wasn't the only one impressed by Ball's competency; her research also attracted the interest of Dr Harry Hollman who worked at Kalihi Hospital. He had been researching treatments for leprosy; it seemed that injections of chaulmoogra oil had been attempted as early as 1888 and there was also some literature on different oils that could be mixed with chaulmoogra extract.[2] Hollman was ready to take this research to the next level and with her knowledge on the chemical make-up of plants, Ball seemed like a good candidate to be his assistant. However, she proved to be much more than just an assistant.

Chaulmoogra oil had been known about for centuries and was introduced as a treatment in Hawaii decades before Ball encountered it.[3] However, applying it to the skin's surface did not show consistent improvement and ingesting it often caused vomiting. As Hollman found during his research, injections of the oil had also been attempted, but the oil was not water-soluble due to its chemical structure. This caused issues when it came into contact with body fluids under the skin. In short, the world had been waiting for a good chemist to come along and find a way to make this oil injectable. Enter Alice Ball. In just a year, she was able to understand the composition of the oil and hence how to separate it into useable and waste parts.

Despite the incredible breakthrough that Ball made, she did not receive proper credit for her work. The president of Hawaii University took her work and expanded on it, calling the extraction process the 'Dean Method' after himself. Having tragically died in 1916 at the age of 24, likely from chlorine inhalation during a laboratory accident, Ball's lack of testing and development was not for want of trying. Science is almost always a team effort, however, what is saddening is that Ball was not given appropriate credit in the published works that detailed her discovery. 'The Fractionation of Chaulmoogra Oil', initially published in the *Journal of the American Chemical Society* and reprinted in the Public Health Reports (1896–1970), only mentions the names Arthur L. Dean and Richard Wrenshall.[4] Similarly, when reviewing articles and interviews for a library exhibit, Paul Wermager, head of Hamilton Library's Science and Technology Reference Department, could not find any evidence of Dean giving credit to Ball's initial discovery.[5]

Luckily, Dr Hollman's 1922 publication provides us with the words that confirm Ball's role in taking the first step towards this novel treatment, explicitly stating that, 'Miss Ball solved the problem for me'. He then goes on to name the method of making specific chemical compounds of the fatty acids in chaulmoogra oil as the 'Ball Method' and even critiques the additional step developed by Dean and his colleague. He stated that it offered no improvement on Ball's original method and added equipment that might not be accessible to all.[6]

As a result of Ball's year of experimentation, many patients who had been hospitalised with leprosy were discharged. The social stigma of being 'unclean' was gradually improved and many dreadfully lonely people, who had suffered years of isolation, were able to reunite with their families.

Tu Youyou (1930–)
Pharmaceutical Chemistry
China

When Tu Youyou contracted tuberculosis at the age of 16, causing her to miss two years of school, winning a Nobel Prize was probably the last thing on her mind. However, this experience left Tu with an interest in clinical medicine and led to her discovering a new way to treat a disease that has plagued humanity for tens of thousands of years.[1]

Born on 30 December 1930, in Ningbo, China, Tu was the only daughter of her parents' five children. Her parents believed in the importance of education regardless of sex, so Tu grew up without the desire to conform to traditional gender roles. With her heart set on staying healthy and helping others to do the same, she applied to Peking University Medical School and was accepted by the Department of Pharmacy in 1951. Tu received a broad education whilst at university and learned about different areas of pharmaceutical science. Although she did not specialise in it at the time, the field of phytochemistry caught her attention and the methods she learned of extracting and characterising the active ingredients from plants would later be the basis of the research that would make her name in the global scientific community.[2] By the time she graduated in 1955, the Chinese government was increasing the focus on combining traditional Chinese medicine (TCM) with Western medical practices. The thousands of existing TCM practitioners were requested to attend training programmes on modern Western medicine and Tu found herself in the other group of newly qualified doctors who were required to learn traditional Chinese medical theory. We are sometimes guilty of thinking that the word 'modern' is synonymous with 'good' but as Tu found out, there was some important wisdom from past generations that could work alongside Western scientific techniques.

In the 1950s and 1960s, scientists realised that existing antimalarial drugs were losing their effectiveness as the microorganisms that caused the diseases were developing resistance. A novel medicine was desperately needed and Tu and her colleagues comprised one of the research groups studying new possibilities.

Her involvement with this line of work started during the Vietnam War. In their conflict with South Vietnamese and US Forces, leaders of North Vietnam found that they were losing soldiers to malaria at an alarming rate as they travelled down the Ho Chi Minh Trail. They asked the Communist Party of China to help provide a solution that would help them lower the death toll and Chairman Mao, whose people were all too familiar with malaria epidemics, obliged. With this agreement to collaborate, Project 523 was born. It was highly secretive; as the disease was affecting the war effort on both sides, any scientific advancement would provide a huge benefit and the Americans already had a research institute looking for compounds that might lead to a breakthrough.[3] The covert Chinese project was based on multiple methodologies. As well as using Western methods of pharmacy to drive novel drug development, there was another group of scientists who were instructed to base their research on TCM.[4] Tu Youyou was chosen to head up this second group at the Institute of Chinese Materia Medica and they quickly got to work, collecting over 2,000 medicinal items over the course of a few months[5] so they could scientifically screen each one in the hope that some might yield useful properties.

Project 523 took place during the Cultural Revolution, during which scientists were perceived as one of the nine 'undesirable' groups that needed to be stamped out in accordance with the aggressive socio-political purge led by China's leaders. The antimalaria project, which had been instigated by Zhou Enlai, the Premier, provided some respite for the several hundred scientists who were recruited. Tu's team was able to provide some world-changing results despite the turbulent political environment they found themselves in. After a painstaking process of trial and error, they found that an extract of the plant *Artemisia annua L.* showed some potential of reducing the growth of microorganisms.[6] Also known as sweet wormwood, this plant can be found in Chinese medical literature dating back to 168 BC and has been recommended for a range of ailments, from haemorrhoids to fever.[7] It was here that Tu's team hit a snag. Even after their careful study of the historical texts available to them, they could not seem to repeat the experiment and achieve the same results. The golden phrase for the team had come from Zhouhou Beiji Fang, *The Handbook of Prescriptions for Emergencies*, by a Chinese physician called Ge Hong who had been writing in the fourth century. The recipe is vague to say the least; his instructions tell the reader to take one bunch of *qinghosu*, soak it in 0.4 litres of water, wring it out, take the juice, and drink it all.[8] The team's problem was eventually solved as they realised that they needed to extract the active ingredient at a lower temperature as the conventional method they were using was destroying the antimalarial properties. Finally overcoming this hurdle, they managed to successfully treat mice and monkeys who had malaria.[9]

Although Tu Youyou and her fellow scientists working on Project 523 were safe from the persecution that the Cultural Revolution encouraged, clinical trials were one of the processes severely disrupted. The development of the much-anticipated antimalarial drug was in danger of being stalled. Tu and her colleagues volunteered themselves to be the first human test subjects. Following a positive result, the drug was then tested on malaria patients. This was even more promising, as the patients who took the *A annua* extract, as opposed to the existing chloroquine medication, showed reduced symptoms. With enough evidence that the extract was clinically safe and effective, the next step was to synthesise a drug from the natural molecule. Towards this end, Tu received help from fellow Project 523 researchers at the Yunnan Institute of Pharmacology. The scientists in Yunnan were instrumental in showing that the source of the plant was linked to the yield of artemisinin, the active ingredient. Once Tu's team replaced the dried leaves they had been using with the more effective fresh leaves recommended by their colleagues in Yunnan, there was a real breakthrough. It was then determined that capsules, rather than tablets, were the most effective drug delivery system, and the newly established drug was used for the first time on soldiers fighting in the Sino-Vietnamese War. The same year, their important achievement was recognised when a National Invention Certificate was presented by the China National Committee of Science and Technology. The Cultural Revolution had lasted just over ten years and ended in October 1976. Gradually, the scientific community began to recover; publishing restrictions were eventually lifted and in 1981 Beijing hosted a scientific working group sponsored by international organisations such as the World Health Organization, the United Nations Development Programme, and the World Bank. Tu spoke about the findings of Project 523 at this event and it wasn't long before she and her colleagues gained a lot of attention. However, her work on the project had demanded her undivided attention so, after sharing her research with the world, she turned her attention back to her family. Her infant daughter had been entrusted to the care of her parents and her 4-year-old daughter had lived with her teacher when not attending nursery. Tu recalls that her youngest daughter did not recognise her during the first visit after years apart.[10] Her scientific discovery had come at a high personal cost.

As Tu herself stated, there is still much research to be done on antimalarial drugs. Carter and Mendis' review on the burden of malaria highlights the disproportionate number of cases in certain African countries. However, the same review shows highly encouraging figures for the rest of the world. The approximate percentage of all deaths due to malaria from the years 1900 to 1997 have decreased from 0.8 per cent to 0.0001 per cent in Europe and North America; from 2 per cent to 0.05 per cent in the Caribbean, Central and South America; and 9 per cent to 0.1 per cent in Asia, China, and the West Pacific.[11] Tu received a share of the 2015 Nobel Prize in Physiology or Medicine. It is fitting that the woman who contributed so much to the incredible change in these figures should go down in the history books as a Nobel laureate, proving to girls and women across the world that it is not necessary to receive a Western education to do great science.

Françoise Barré-Sinoussi (1947–)
Virology
France

Not only a talented researcher and the recipient of a Nobel Prize in Physiology or Medicine, Françoise Barré-Sinoussi's many awards and involvement with research programmes and organisations show her commitment to linking education, action, and research to enable science to benefit the global population.

Born in July 1947 in Paris, Barré-Sinoussi showed both enthusiasm and scientific talent from an early age. She chose to study natural sciences at university, not wanting to inconvenience her family with paying for what she thought would be a longer and more expensive medical degree.[1] After a long search for a laboratory to volunteer at during the latter part of her degree, Barré-Sinoussi received an offer from the Institut Pasteur site at Marne-la-Coquette, where Jean-Claude Chermann and his research group were studying the link between retroviruses and cancer.[2] She became engrossed in her work and, before long, Chermann suggested she do a PhD.

Genetic information is conveyed by the molecule messenger ribonucleic acid (mRNA), which is formed when DNA undergoes a process called transcription in the nucleus of a cell. Reverse transcription is the opposite of this; retroviruses operate in this way by inserting their own RNA into the host cell DNA. During her PhD research, Barré-Sinoussi studied a type of leukaemia in mice that was caused by this type of virus. She experimented with artificially inhibiting the enzyme that facilitates the reverse transcription of RNA to DNA, thus preventing the virus from interfering with the genetic information in the host cell. In 1974, she was awarded a PhD by the Faculty of Sciences at the University of Paris.[3]

Having expanded her network with two researchers who had spent part of their sabbaticals visiting Barré-Sinoussi's group, she accepted a post-doctoral fellowship at the National Institutes of Health in the United States. She spent a year abroad looking at how the replication of the leukaemia virus in mice is genetically restricted, before returning to France as she had been offered a role at the Institut National de la Santé et de la Recherche Médicale (INSERM). This meant that she was able to once again work in Jean-Claude Chermann's laboratory, this time in the unit headed up by Professor Luc Montagnier.[4]

Little did Barré-Sinoussi know at the time, the knowledge of retroviruses that she and her group had been building up was about to become extremely important. The 5th June edition of the 1981 *Morbidity and Mortality Weekly Report* was to be the starting point of the race against an epidemic that would go down in history for the discrimination, stigmatism, homophobia, and racism that it caused. The report detailed the pneumonia diagnosis of five males over a six-month period in California, US.[5] What made this stand out was that they all had a form of pneumonia that was only found in people with very weak immune systems; as all five men had been healthy prior to diagnosis, this did not make sense. The patients all died from their illness and, alarmingly, reports of similar deaths became more and more common and started to include other opportunistic infections such as Kaposi's Sarcoma,[6] a type of cancer also linked to compromised immune systems. Initially, only gay men were being diagnosed, which led to the incorrect conclusion that the disease could only affect gay people. There are sources that show it was unofficially referred to as gay-related immunodeficiency disorder (GRID)[7] which was not only misleading but added to the prejudice that much of the LGBTQ+ community already experienced. A year after the West became aware of this disease that would later become known as acquired immune deficiency syndrome (AIDS), cases of heterosexual women being infected started to appear.[8] Other groups that seemed disproportionately affected were heroin users, haemophiliacs,[9] and people from Haiti. As the epidemic progressed and public hysteria surrounding the mysterious condition grew, the pressure was on to find the cause of this global health crisis. The only known human retrovirus at the time was human T-cell leukaemia virus (HTLV), and Barré-Sinoussi and her colleagues at the Pasteur Institute were asked whether this could be causing the epidemic. These retrovirus experts noticed one key difference: patients with AIDS seemed to have a decreased number of T helper cells, a type of white blood cell that has a vital role in the specific immune response by binding with antigens from pathogens. HTLV was not known to cause the depletion of these cells,[10] which suggested that it was not the retrovirus that caused AIDS. In early 1983, Barré-Sinoussi and Montagnier started an experiment, using the hypothesis that a new human retrovirus existed. They extracted a sample from the lymph nodes of a patient showing symptoms of AIDS and measured the activity of the enzyme, reverse transcriptase, over a period of time. The activity increased, peaked, then decreased as the host cells died.[11] Despite initial concerns that cell death was occurring due to issues with the tissue culture, the result after adding more lymphocytes confirmed that cell death was caused solely by the virus. Confident that the virus they had discovered was not the same as HTLV, they wrote a report on what they named lymphadenopathy-associated virus (LAV) that May.[12] From this point, a large effort was made to focus research on further characterising the virus. Similar reports of a novel retrovirus submitted by an American research group turned out to be based on a sample sent to them by the Pasteur Institute, so Barré-Sinoussi and her colleagues were able to take credit for discovering the virus that is now called HIV.[13] They managed to convince the scientific community that it was linked to AIDS and this marked the beginning of an international collaboration between researchers, clinicians, charities, and governments to develop and distribute treatments as well as educate on preventative measures.

For her incredible efforts, Barré-Sinoussi shared the 2008 Nobel Prize in Physiology or Medicine. She continued research work until her mandatory retirement in 2015,[14] as well as having heavy involvement in international outreach activities. The AIDS epidemic continues to this day and continued research into more effective treatments and vaccines is vital, given that figures from 2019 show an estimated 38 million people were living with HIV, with 690,000 dying from AIDS-related illnesses in that same year.[15] These numbers are significant; however, the identification of HIV was the event that all subsequent research was built on and a major reason why AIDS is now very preventable and treatable. With access to the right treatment, HIV positive individuals can live healthy lives and for this reason, Barré-Sinoussi's contribution to STEM and the global population was great indeed.

Rosalyn Yalow (1921–2011)
Medical Physics
United States

Rosalyn Yalow (née Sussman) was a traditional woman in many senses, and placed great importance on the traditional role of women in the household. She also happened to be a talented medical physicist and recipient of a Nobel Prize.

Yalow was born in New York in 1921, to parents who came from eastern European immigrant families. Though not educated beyond a primary education themselves, they were determined for their children to have the opportunity to go to college. After completing high school in the Bronx area of New York, Yalow attended Hunter College, which started offering a physics major just in time for Yalow to specialise.[1] It was a female-only institution at the time, and things became more difficult once Yalow started to plan for her future. Despite being encouraged by her family to pursue a career in teaching, she had her heart set on physics, making her options somewhat more limited. One particular institution insinuated that both her gender and Jewish background would be a barrier to future employment and that it was not willing to consider her on those grounds.[2]

When the US entered into the Second World War, the demographic of higher education changed dramatically as students and staff were required to offer their services to the war effort. As male applications dropped, the University of Illinois at Urbana-Champaign was one institution that started admitting women to their courses,[3] and Yalow was successful in getting a position as a teaching assistant in the physics department. She was able to sit in on graduate courses as well as undergraduate courses for topics she wanted to improve her knowledge of. Receiving her master's degree in 1942, she married fellow physics student Aaron Yalow the following year and earned her PhD in nuclear physics in 1945. Whereas her husband would go on to have a career in teaching, Yalow was drawn to research. Whilst she did take a teaching job at Hunter College back in New York, she spent her spare time learning about how radioisotopes could be used for medical purposes. This was an emerging field at the time and Yalow was recommended to the chief of radiotherapy at the Bronx Veterans Administration Hospital (also known as Bronx Veterans Hospital and currently the James J. Peters VA Medical Center) as he was interested in developing their new radioisotope service that existed in name but not in function.[4] Yalow was just the woman they needed; she had spent much of her doctoral research building apparatus that could measure radioactive substances. There was no doubt about her initiative and creativity once she set up a functioning service inside a caretaker's cupboard.[5] She and her colleagues published several papers even though she was still teaching full-time up until 1950. Taking the decision to focus solely on research was a pivotal point in Yalow's life as this was also the year she met the man who would become such a part of her work that their relationship was often referred to as a scientific marriage. Dr Solomon Berson was a medical doctor who equalled Yalow in her strong-willed, ambitious, and sometimes aggressive temperament. Like their personalities, their skill sets aligned beautifully. Yalow taught Berson physics, and Berson was such a comprehensive teacher that his colleague never needed to have any formal biology training, despite their research being centred on the human body. Radioactive isotopes had been used to treat cancer, but Yalow and Berson understood that there was a broader range of applications. By monitoring the rate at which a radioactive tracer decayed, they determined blood volume and expanded this method to improve diagnosis of thyroid diseases.[6]

Their next scientific investigation was closer to home as Yalow's husband was diabetic, but the understanding of the condition was not fully developed at the time. The clinical scientist I. Arthur Mirsky had a theory that type II diabetes might not be caused by a deficiency of insulin but the degradation of existing insulin.[7] Type II diabetes occurs when a person can produce insulin but retains a high blood sugar level because the insulin they produce cannot remove the sugar. Yalow and

Berson found that when they injected pig and cattle insulin into patients, antibodies were produced as part of an immune response.[8] This was a breakthrough, as prior to these results it was understood that the molecular size of insulin was too small to initiate an immune response. As can happen, their results were so surprising that they had difficulty convincing the scientific community that they were correct. Despite the initial cynicism, Yalow and Berson published their paper explaining how the reaction between an antibody and an antigen could be studied using radioisotopes.[9] The next logical step would be to say that by measuring the antibodies that had bound to insulin, the amount of insulin is able to be determined. They called this technique radioimmunoassay (RIA) and it changed health screening forever. Not only is RIA sensitive enough to pick up on very small concentrations of proteins but it is cost effective and can test multiple samples at once. In the years that followed, Yalow and Berson showed that by using RIA to analyse blood samples they could also measure tumour antigens, vitamins, and enzymes. This was a vital discovery as it meant that conditions in newborn babies and young children could be diagnosed before the symptoms started to show or irreversible damage occurred. Yalow and Berson also used RIA to measure levels of the virus Hepatitis B; this was hugely significant as blood banks now use the same technique to test for the virus before any donated blood can be used.[10] RIA had huge commercial potential but Yalow and Berson made the selfless decision not to patent the technique and went to great efforts to promote its use, even arranging courses to share their knowledge and train medical professionals.

Berson was old-fashioned and, like many men of his time, displayed traits of male superiority;[11] however, he was fair when it came to publications and apportioning credit. Yalow and Berson alternated first authorship on their research papers even though a female and male team at the time would lead many outsiders to believe that the woman played more of an assistant role.[12] Perhaps it was because of this preconception that Berson's premature death in 1972 affected Yalow professionally as well as personally. He had moved on from their partnership in the late 1960s, but there were those who incorrectly believed that Berson was the intellectual driving force behind their successful work on RIA.[13] Yalow paid homage to her late friend and colleague by naming their laboratory after him, and carried on their work with her new partner, Eugene Straus.

During this time, she proved just how productive she could be. Yalow and Straus published around sixty papers and in 1977 it was announced that she was to become a Nobel laureate, making her the first US-born woman to be awarded a Nobel Prize in a scientific field. Her success was honoured in other ways as well: Yalow became part of the US National Academy of Sciences and was the first female president of the Endocrine Society. [14] Women's awards however, were of little interest to her. Notably, she turned down a *Ladies Home Journal* 'Woman of the Year' award as she did not understand the relevance of gender to her work, explaining that it was 'unwise' for celebrated achievements in a field as broad as science to be restricted to one gender or another.[15] Yalow continued to go to the laboratory after her retirement, despite suffering a number of strokes.

Historically, the sacrifice of marriage and/or children was often the required price for a woman to have a career in STEM, but Yalow believed that women should also be able to take the lead domestically.[16] Her life reflected more modern-day ideals; she put just as much effort into her family and home as she did in her laboratory. Whilst the Yalows were fortunate enough to have a full-time nanny whilst their two children were young, she took on all cooking and household chores in later years. The range of modern-day feminist viewpoints have been shaped over time by multiple factors that include sexism, racial discrimination, and environmental issues. By the time Yalow died in 2011, a lot of positive improvements had been made for women in some industries. However, until there is a widespread and consistent shift in workplace culture across the globe, the challenges of balancing work life and family life are likely to continue for women in particular. The feminist movement has quite rightly grown more complex than campaigning for more equal treatment of niche subsets of the female demographic, yet Yalow remains an important role model for women who value aspects of traditional gender roles whilst aspiring to have a successful career in science.

Chapter 5

Protecting the Earth

Saruhashi Katsuko (1920–2007)
Geochemistry
Japan

How acidic is rain, and has it changed over time? Can we measure how much carbon dioxide is in the ocean? What is the environmental impact after a nuclear explosion?

Saruhashi Katsuko cared deeply for the earth and its people, and used science as a tool to answer these questions. The discussion about global warming and CO_2 emissions has become widespread in recent years. Both households and industry are being encouraged to reduce their carbon footprint in order to reach the global sustainability targets that are vital for protecting the future of the planet and every living thing on it. Saruhashi was born in Tokyo, Japan in 1920, at a time when public awareness was very different.

Although she had many options available to her, Saruhashi decided to study science at what is now the Toho University. This decision may have been influenced by her observing the number of widows and fatherless women resulting from the horrors of the Second World War, and experiencing the damage that could be achieved by scientific innovation.[1] Using science for the 'happiness of humankind' was just as important to her as financial independence, which she thought would be more achievable by finding a technical skill.[2]

After earning her degree, she started working at the Meteorological Research Institute, using chemistry as a foundation to study important geochemical issues. In 1957, she earned a PhD from Tokyo University, becoming the institution's first woman to achieve this level of qualification in a scientific discipline. Around this time, her work involved measuring and understanding the carbon dioxide content of seawater; until that point, this was relatively unexplored and under-researched. Due to previous studies, it was thought that the increase in CO_2 levels in the ocean were mainly caused by a compound called calcium carbonate as it dissolves into the water. Saruhashi and her mentor, Yasuo Miyake, updated this theory by showing that the increase was actually due to the oxidation of organic matter.[3] Additionally, the paper she published in 1955 described an improved method of calculating accurate levels of carbonic acid (carbon dioxide that has reacted with water) by also measuring temperature and pH.[4] She even acknowledged that the calculations she proposed were time-consuming and as a result, included what is now known as the Saruhashi Table;[5] this was a time before computers, and the data she provided would prove invaluable to oceanographers and fellow geochemists. This accurate method makes Saruhashi a pioneer in understanding the natural world, which is so important when attempting to track and monitor changes to the planet caused by increased environmental pollution. Her work involving the Pacific Ocean's release and absorption of carbon dioxide also went on to discourage the idea that CO_2 can be stored within the ocean.

Over 2,000km south-east of Tokyo, there was activity happening that made chemical analysis of the ocean not just interesting but vital. Since the end of the Second World War, the US had been conducting nuclear explosions in remote locations in the South Pacific, including the Marshall Islands and Bikini Atoll. By 1958, a total of sixty-seven tests had been carried out.[6] Worryingly, the full impact on the environment and health of living beings was not understood at this point. Although locals had been displaced to reduce the short-term harm caused to human life, the Japanese government went a step further and asked Saruhashi and her colleagues to investigate the long-term and wider-reaching impact of the bomb test sites. They did this by conducting a research project and publishing their findings in a paper, simply titled 'Cesium 137 and Strontium 90 in Sea Water'. The fears of the Japanese government were confirmed. Saruhashi and her team reported that these two radioactive substances were much more concentrated in the western North Pacific than in the Atlantic and eastern North Pacific.[7] As recent disasters involving radiation have shown, radioactive contamination of oceans and soil is dangerous to life and difficult to avoid without resorting to radical measures. The research Saruhashi's team conducted contributed to the scientific evidence that provided such a significant warning that in 1962, the US halted all nuclear explosive tests. By the following year, the Test Ban Treaty had been signed in triplicate by the US, the UK, and the

Soviet Union,[8] which was incredibly meaningful given the ongoing threat of the Cold War. In due course, 105 other countries would go on to sign a form of this treaty. This was possible in part because Saruhashi gave her support to a knowledge exchange with an American counterpart who had been given the same task of determining nuclear fallout from the test sites. She worked at the Scripps Institution of Oceanography (SIO) at the University of California, San Diego for six months in 1962. In Sumiko Hatakeyama's biography of Saruhashi, she writes that the scientist who was talented enough to be trusted with a government-funded research project was instructed by her hosts to work in a wooden hut rather than use the laboratory facilities like her peers.[9] With a welcome like that, it must have seemed like a long half-year but perhaps the change in legislation that it resulted in seemed worth the sacrifice in the end.

Saruhashi was fortunate in the opportunities she was given, the mentors she had, and the lack of gender discrimination she was subjected to in the workplace. However, she never took this for granted and did great work mentoring and providing safe spaces for female scientists to gather together and discuss challenges facing women working in STEM fields. She founded the Saruhashi Prize in 1981 to reward female scientists and provide money that could be put towards further research and study. In Japanese, the name 'Katsuko' translates to 'the one who wins/is victorious'. Thanks to her determination and hard work, this Japanese pioneer of STEM certainly lived up to her name.

Rachel Carson (1907–1964)
Marine Biology
United States

A scientist, writer, and passionate conservationist among many other things, Rachel Carson went to extreme lengths to uncover harmful practices and educate the public on the wonder of nature, as well as the human capacity to cause irreversible damage. She used her writing skill to eloquently describe the issues facing the planet's future, took her cause to the US senate, and even hid her illness when she developed cancer so as not to appear biased to the chemical companies she was standing up to.

Carson was born in Pennsylvania, US, in 1907.[1] She did not come from a wealthy background, but she and her siblings had a childhood rich in learning and surrounded by nature. Carson's mother encouraged her to develop a good understanding of the natural world around her: the plants and animals, as well as how they interacted with the environment and each other. This upbringing was the real foundation of Carson's later work. Whenever she faced opposition from intimidating corporations, whenever experts debated the issues she had brought into the public interest, deep inside, Carson must have had quiet confidence that she truly understood the beauty of the natural world and the consequences of damaging it beyond repair.

Carson's mother was keen for her daughter to attend college after she left school. Finding the money to accommodate this was not easy, but with the help of a scholarship, Carson was able to receive the education that would allow her to fulfil her potential. She had enjoyed writing since childhood, and her work had been of high enough quality to be published in several magazines. Her outdoor explorations provided inspiration for much of what she wrote, so the fiction she was encouraged to write as part of her course at Pennsylvania College for Women proved to be somewhat of a challenge. Fortunately, Carson would imminently find a role model who would not only provide inspiration but support her career for years to come. Mary Scott Skinker became Carson's biology tutor in 1926 and then her mentor when she decided to change her major to biology a couple of years later.[2]

Carson graduated in 1929 and completed a summer scholarship in Woods Hole, Massachusetts. Here at the US Marine Laboratory, her childhood fascination with the ocean became serious. She was only able to go diving once during her time studying marine biology, due to superstitions surrounding women being onboard.[3] Despite the challenge of accessing practical experience her continued interest led her to do a master's degree in zoology at Johns Hopkins University.

By the time Carson finished her second degree in the summer of 1932, the US was well into the Great Depression and the job market was bleak. Any ambitions to complete a PhD were dashed by her family's financial situation; Carson's parents would welcome any income she was able to provide, even more so after her father's sudden death. Despite her qualifications and experience, Carson had to settle for a part-time position at the US Bureau of Fisheries where she was tasked with writing educational broadcasts, a role she excelled at.[4] Her ability to engage the general public with the fish biology of the local area impressed her employers, and her career path changed direction the following year when more job vacancies became available. Her former tutor, Skinker, who by this time had helped Carson receive the funding for her time in Woods Hole and endorsed her application for her master's degree, offered her assistance once again by providing a recommendation and preparing her for the federal civil service exam.[5] The combination of this careful mentoring, along with Carson's innate intelligence, led to her passing the necessary exams with flying colours and being offered the position of junior aquatic biologist in 1936.[6]

By now, Carson was building her experience and knowledge of aquatic life at a steady pace. Writing for the radio programme *Romance Under the Waters* had provided plenty of research material and she was so taken with her findings that she frequently wrote articles and submitted them to newspapers. Her writings outside of the office eventually led to her first book, *Under the Sea Wind*, which was published in 1941. In this debut work, Carson's mastery of the English language and story-telling blends beautifully with her keen sense of observation. The tone is excellently placed, somewhere between a nature documentary and a children's book; the way in which she introduces the three animal 'protagonists' and shows how their lives are not independent from other species or their environment is completely accessible to her readers. This was reflected in the good reviews received; however, the book was launched around the time of the Japanese attack on Pearl Harbor.[7] Public interest was engaged with the war and the book did not sell well.

Back at her day job, Carson had shown her male-dominated workplace that her gender had no negative impact on what she was capable of. She received a promotion in 1943 at the newly established US Fish and Wildlife Service,[8] later becoming chief editor of all publications. Throughout her fifteen years as a government employee, her roles built on her pre-existing talent of explaining science in interesting yet understandable ways. Additionally, she had the opportunity to visit plants and talk to people whose daily work was affected by the waterways. This provided her with an understanding of the economics of the area as well as the ecology and conservation requirements; a balance that would prove vital as she went on to become a more outspoken environmentalist.

Given the tepid reception of *Under the Sea Wind*, Carson waited another ten years before publishing her next book, *The Sea Around Us*. In contrast, her sequel became a bestseller as well as making her a respected household name and authority on science for a post-war population. This was a turning point for Carson in more than one way. For her first two books, her role as a government employee required her to be careful with her tone and opinions. Following the success of *The Sea Around Us*, Carson was able to hand in her notice, and along with this disappeared the need to be passive.

This was timely as trouble was brewing in the agricultural world and Carson soon unwillingly found herself at the heart of it. During the Second World War, the chemical Dichlorodiphenyltrichloroethane (DDT) had been used to control the spread of insect-borne diseases.[9] It was thought to be so effective that the chemical was sold in the US as a household and agricultural pesticide shortly after the end of the war. The uptake was swift and widespread.

Gradually, however, people started to notice some unwelcome changes in their environment. As a well-known figure and conservationist, Carson's desk was soon piled high with the anxious concerns of fellow citizens. Unable to find anyone else to write about the overuse and potential consequences of harsh chemical pesticides, she started to compile information and correspondence from people involved with and affected by the use of chemicals being sprayed up and down the country.[10] This resulted in Carson's famous book, *Silent Spring*, published in 1962. Going against the establishment, including some scientists, meant that the data and sources Carson built her book on had to be watertight. Government employees would not have been permitted to share information about their work but somehow, some information found itself into Carson's hands. She was careful not to sensationalise the issue or demand cessation of all pesticide usage, but what she did provide was a warning that actions we take to harm the biodiversity and ecosystems of our planet do not happen without consequences. There was little research to show the effects of these chemicals on human health or surrounding wildlife, and there was little logic as to why the extensive use of DDT had been encouraged so hastily by both government and industry. Ultimately in her writing, Carson condemns the idea of humans attempting to exert full control over nature and she uses emotive language to suggest that the power dynamic has shifted to become 'humans against the earth', rather than a more sustainable symbiotic relationship.[11] *Silent Spring* caused outrage, and the inevitable backlash came crashing down on Carson as her powerful words awakened the public. She stood her ground, even when she was taken to court by large chemical companies, and did so whilst severely weakened by cancer.[12] She died at the age of 56, a year after giving her testimony to John F. Kennedy's Scientific Advisory Committee. The report issued by the committee was in agreement with many of Carson's claims. In 1980, she was posthumously awarded the Presidential Medal of Freedom. And her influence did not end there. Her writings provoked government action and led to the following policies.

- Clean Air Act (1963)
- Wilderness Act (1964)
- National Environmental Policy Act (1969)
- Clean Water Act (1972)
- Endangered Species Act (1972)
- Establishment of the Environmental Protection Agency (1970)

Exploiting the Earth's resources with little consideration for the environment, sustainability, and human life is a quick and profitable way to drive what many see as the advancement of nations and economies. The life-long work of Rachel Carson confronted the world with the hard truth of what we are willing to pay in the name of 'progress' and her work was a catalyst for the modern environmental movement.

Further Reading

Mary Somerville – a Scottish scientist and writer who was jointly the first woman to be accepted into the Royal Astronomical Society in 1835. She had a talent for science communication and published on a range of subjects from physical geography to classical physics. Somerville played a large role in popularising science and she was well respected by the public as well as within her own social circles. She was the tutor of Ada Lovelace (p.6).

Wangari Maathai (1940–2011)
Veterinary Anatomy, Environmental Science
Kenya

Born in Kenya, Africa, in 1940,[1] Wangari Maathai (née Muta) was a role model for those aiming to take a more active role in the environmental movement. She had a deep understanding of how to tailor discussions on sustainability to her audience, whether it be addressing the UN, debating with members of parliament, or encouraging local people to get involved with the grassroots organisation she set up to combat deforestation. Also the first woman from East and Central Africa to earn a PhD and the first African woman to receive the Nobel Peace Prize,[2] Maathai showed the world that background and family wealth should not hold girls back in their aspirations to access an education in STEM.

Maathai grew up in Kenya whilst it was a British colony. After moving around as a young child, she spent several years attending St Cecelia's Intermediate Boarding School where she was able to avoid the worst of the Mau Mau uprising against the British authorities that gripped the country between 1952 and 1960. Her mother was not so lucky and had no choice but to evacuate to an emergency village in Ihithe, Maathai's birthplace. This period of violence was the result of decades of systemic injustice inflicted on the native population during colonial rule. There had been resistance to British imperialism ever since the country we now know as Kenya was established as a protectorate in 1895 and a colony in 1920, but the Mau Mau uprising started in earnest when internal divisions lead to some factions employing more violent tactics, which the British met with brute force. This became the longest period of anti-colonial warfare in Kenya. The estimated death toll ranged from 11,000 to 25,000;[3] although the British put an end to the revolt, Kenya gained independence in 1963.

Because of her good grades, Maathai was able to get a place at Loreto High School in Limurus, followed by a place to study biology with chemistry and German at Mount St Scholastica College, Kansas. This was made possible due to the vision of a young Kenyan leader called Tom Mboya, who believed that offering the opportunity for the brightest students to study abroad would help prepare the nation for a post-colonial Kenya after the European civil servants left. To achieve this goal, from 1959 Mboya travelled the US extensively, asking colleges to provide scholarships. Combined with the fundraising efforts of the students' families and tribal groups, and the African American Students Foundation (AASF) which had been founded by several influential public figures, financial provision for living and travel costs was secured. By the time Maathai's cohort of 300 students were due to travel to the US in the autumn of 1960, the programme was in dire need of funding to cover air fares. John F. Kennedy, who was a senator at the time, agreed to provide $100,000 towards this cause, which is the reason that this period of the programme was nicknamed 'The Kennedy Airlift'. In total, over 800 students[4] benefited from the educational scheme. The scholarship programme officially ended in 1963, but when Maathai earned her bachelor's degree a year after this, she received funding from the Africa-America Institute and went on to complete a master's degree in biology at the University of Pittsburgh. It was during this time that Maathai became aware of environmentalists working to tackle pollution in the city; the restoration she observed may well have sown the seeds of her future passion for environmental justice. Like many of her peers who had accessed scholarships in the US, Maathai returned to Kenya after graduating. She was successful in obtaining her first graduate job as a zoology research assistant at what is now the University of Nairobi, but unfortunately the post was given to someone else before she was able to start. After months of looking for another job, she found an opportunity with Professor Reinhold Hofmann, who was based at the University of Giessen in Germany but was willing to hire her to work at the Department of Veterinary Anatomy at the University of Nairobi. This was the start of a new chapter for Maathai as her employer would later encourage her to do a PhD, which would bring with it the opportunity to work abroad; whilst working towards her doctorate she studied at both the universities of Giessen and Munich in Germany. Other influences were also afoot around this time: she met Mwangi Mathai, the man who would become her husband and introduce her to the world of politics through his own career. They married in 1969 and after completing her doctorate in 1971, Maathai started lecturing at the university and simultaneously helping her husband in his campaign to be elected as a member of parliament for the Lang'ata constituency.[5] One thing was abundantly clear: people's main concern was job security. The campaigners made promises that they would improve opportunities and Mwangi Mathai was elected. His wife went about trying to keep the promises they had made to the people of Lang'ata. She had a big heart, and worked on a solution long after the issue had been pushed to the side by politicians. With the help of a family friend and the forester who looked after Karura forest, Maathai set up her plan to start a forest nursery that would employ local people to plant and care for the trees. She also founded her company, Envirocare, from their home to enforce the reforestation effort. In what was the first of many trials throughout her life, Maathai gradually realised that Envirocare was not taking off at all. The locals she had hoped to employ were so destitute that they required financial dispensation that she was not able to give, such as transport money and payment upfront. Even on a broader scale, the interest the company had drummed up during the International Show held in Nairobi in 1975 seemed to go nowhere. By this time, she had three children. Managing family life as well as teaching and developing a business meant a full-on schedule, but Maathai's friendly and caring nature had led to her building up a good network and her fortunes started to change for the better in 1976. At her university position, she was promoted to Chair of the Department of Veterinary Anatomy and went on to become an associate professor. As for her passion project, she was invited to take part in the first conference on human settlements (HABITAT I) which was held by the United Nations in Canada. Seeing how many people from across the globe were concerned about environmental issues and dedicated to finding practical solutions reignited her drive, and Maathai returned to Kenya with renewed vision. Yet once again, luck was against her and a hosepipe ban had made it impossible to keep the tree seedlings alive. Just as it looked as though she had lost everything, she was invited to attend the annual meeting of the National Council of Women of Kenya. This council that had been set up shortly after the country gained independence[6] provided a forum for the many women's groups across the country. Maathai did not see it as a way to publicise her company, but the networking opportunities would eventually give her a platform to talk about the socio-economic and environmental benefits of tree planting.

She took the concept of community-based tree planting even further the following year by setting up the project that became one of her many legacies; the Green Belt Movement. This was later introduced to other countries and went on to have a significant impact. Unlike Envirocare, this project had managed to receive funding from the United Nations Voluntary Fund for Women. This seed fund enabled the movement to grow, for planters to be paid a stipend and upskilled when necessary, and for the work to be recorded and planned efficiently by literate locals. An estimated 45 million trees were planted.[7] This work eventually fed into other organisations; Maathai went on to work with the United National Environment Programme, which smashed their initial target of 1 billion trees, and promptly moved onto a more ambitious target of 14 billion.[8]

Chapter 6

Influential Projects and Leadership in STEM

Kate Gleason (1865–1933)
Manufacturing
United States

If life not going to plan and arrangements falling through sound familiar, Kate Gleason is a rather apt role model.

The story of Gleason's legacy starts before her birth. By 1855, William Gleason, her father, had emigrated to the US from Ireland with his widowed mother and brother, and found a wife, Winifred Lynch. The following year, their son Thomas was born, and their daughter Mary joined the world in 1858.[1] The first of William Gleason's tragedies struck when Mary died aged only 1, closely followed by her mother. Although he had relocated his family to Chicago only two years previously, now a widower, William Gleason was forced to return to Rochester, New York, where his mother and older brother James lived. When James was killed in the Civil War, William Gleason became the head of the household. He married Ellen McDermott the same year and headed to Connecticut in an attempt to find a job as a machinist for a gun manufacturer. He was eventually hired as a mechanic after showing skill with a lathe, and there he stayed until the Civil War came to an end in 1865. In November of the same year, Ellen gave birth to the couple's first child, Catherine – Kate to her parents – around the same time as William Gleason and his friend John Connell decided to open up their own machine shop. Three years later and business was going exceedingly well; his tool rest design had been patented and they had brought in a third partner as well as three employees to aid production. However, it was his career move in the early 1870s that really changed life for his family. Due to a difference of option between himself and the two other partners, William Gleason went to work for Kidd Iron Works who manufactured machines – from lathes and upright drills to steam engines. It wasn't long before he had earned another patent and was asked to design a machine tool that would enable the accurate mass production of bevel gears – an improvement the automotive industry would later greatly benefit from. Just a few years after starting at the firm, William Gleason became the owner.

Meanwhile, his daughter Kate was growing up in a man's world. She had always been close to her half-brother Tom, but by this time she had two younger brothers, James and Andrew. This was probably the reason why Gleason said 'I was trying my best to be as nearly a boy as I could.' In her own words, she added, 'girls in this world were awarded second place, and I resented being second.'[2] Sadly, Gleason's opportunity to prove the world wrong came at great personal loss to the family and also her father's business. In 1876, Tom caught typhoid fever and died suddenly. This was a terribly significant blow to William Gleason; not only had he lost his eldest son and last member of his original family, he had also lost the heir to his company. He was said to have exclaimed to his wife, 'Oh, if only Kate had been a boy!'[3] One can only imagine what the 11-year-old Gleason, who unfortunately heard this outburst, felt upon realising that she was deemed useless for the family business. Being the headstrong girl that she was, her immediate response was to go to her father and demand work. In her article, 'The First Lady of Gearing' in *Gear Technology* magazine, Nancy Bartels muses over whether William Gleason's acceptance of his daughter's offer was due to being distracted by grief, or helping Kate through hers [4]. Whatever the reason, it was an excellent decision, and one that may well have prevented significant deterioration of the company in years to come. Gleason's first job was to sort the bills her father handed her,[5] but later, following a few very successful years, the company was in dire straits due to the banking crisis and she stepped up again, this time as bookkeeper. There was no doubt about Gleason's immediate and unwavering dedication, even at the age of 14. As she later recalled in an interview for an article in *American Magazine*, studying was done from 4 a.m. onwards. School would occupy her time between 8 a.m. and 1 p.m. after which she would have a quick lunch and work until 6 p.m.[6] The value of her role in the business can be clearly seen by the generous salary of $1,100 per annum, offered to her replacement when she left home in her late teens. Thanks to the building of pipelines during the oil boom, business started to recover.

The Civil War had had a dramatic impact on the demographic of the US; women now outnumbered men[7] and this inevitably raised new questions in a variety of sectors, including higher education. The founders of Cornell University had not written a clause regarding male exclusivity into the charter and therefore, at least in theory, any person who could pass the entrance exam should have a place offered to them regardless of gender. The first woman to challenge social norms was Jennie Spencer, when she made a successful application in 1870.[8] Unfortunately, she was unable to graduate given the lack of basic facilities for women, but the graduation of the first female student occurred only a few years later in 1873. The addition of Sage Hall, which offered female accommodation, seemed to encourage applications as the university admitted forty-nine women in the autumn of 1875.[9] By the time Gleason passed the entrance exam, things were progressing in the right direction regarding co-education. This being said, given her subject choice of mechanical engineering, Gleason was still considered an outlier – so much so that she made the news as the only female student in the department.[10] Her classes included shop work with the male students, as well as German, geometry, algebra, instrumental drawing, and trigonometry.[11]

After Gleason moved away from home, her father was optimistic that business was going well, though a letter she received from James in the September of 1884 suggests things were 'as bad as ever'.[12] Confirmation came in writing from her father in May 1885.[13] The situation must have been financially difficult for William, but the strong bond between father and daughter was evident as this letter urges her to finish the term before returning home and that she was to let him know immediately if she needed any money to do so.[14] Gleason wrote that her 'heart broke utterly'[15] upon realising that she had to give up her well-deserved education. Her recollection of despair gives a small insight into this emotional time. She described how she had taken her father's letter and cried whilst seated under a tree, trying to find privacy from her fellow students. The fact that she was the only woman on the engineering course was very important to her and she had been desperate to see it through.[16]

Her father on the other hand, was sure that one year of Gleason's input would restore the business to stability and that she could return to her studies after this period. This dependence on his daughter was a far cry from the days he spent wishing she was male! In any case, Gleason came home and took charge of all business correspondence. She later admitted that a large part of her work ethic in the years that followed dropping out of university was formed by her determination to equal the male peers she had left behind.[17] But there was perhaps more to Gleason's new-found energy than her competitive nature, as she was starting to take interest in things outside of her new role at Gleason Works. The Fortnightly Ignorance Club, led by Dr Sarah Adamson Dolley and Jenny Marsh Parker, met in the offices of their husbands to research, discuss, and plan social reforms for topics that were important to local business women and professionals.[18] Gleason was invited to be a member in the autumn of 1886, and being around these women must have been a vital source of encouragement and inspiration. Only a year before, she had been torn between completing a degree in a subject she loved, and duty to the family business. However, as the years passed and she found herself surrounded by other women who did not comply with societal expectations, formal qualifications may have become less of a priority. As her father had expected, Gleason returned to Cornell University in the spring, but by the end of term she was ready to come home. She even mentioned in a letter to James that she was considering her mother's advice of attending a ladies' finishing school, in order to acquire the 'polish' that may improve customer interactions for the business more than any engineering degree. In this one statement, Gleason proves that she is just as much a business woman as an engineer.

By 1890, Gleason, aged 25, was both secretary and treasurer for Gleason Works and finding that her assessment on the limitations of a formal qualification was correct. The correspondence between her and James continued, though this time, her brother was the one asking after the family business from his student rooms at Cornell. Meanwhile, Gleason was busy being 'office boss of the Gleason Iron Works',[19] and learning skills her degree would not have offered. With her father away on business so much, her management skills were also put to the test. There was an incident in which an employee was bringing whiskey to work; seemingly, Gleason had sufficient authority to order his dismissal if he was caught again. She also spoke excitedly about the four-day 'drumming trip' she went on to Ohio, which included attending an exposition. On the first day of the trip, she managed to secure an order, despite initial worries that the customer would not take kindly to discussing business with a pretty young woman. On top of her work life, Gleason attended a gathering known as 'German Club' during which she had the opportunity to socialise with what she called, 'the cream of German aristocracy'.[20] The termination of her university education was proving not to be the end of her career, more so the beginning.

After several good years, business at the Gleason Works took another turn for the worse in the early 1890s. A fire in 1889 had resulted in the loss of all tool drawings and damage to some of the machines. Thanks to William's tenacity, the tools were designed from scratch, better than before, and the firm lived on. James added further value by automating the bevel gear-cutting machine, and made this the company's primary product. Unfortunately, this was not enough to combat the repeating economic cycle of inflation and depression and by 1893, Gleason was both extremely concerned and physically ill. Presumably to prescribe her some good old-fashioned rest, her doctor advised a holiday to Atlantic City. So, what did Gleason do? She went to Europe instead. Armed with a few hundred dollars, some letters of recommendation, and a single black dress she went on her trip, all the while claiming that any recuperating could be done during the two-month voyage across the Atlantic!

The trip brought about a transformation in Gleason. After returning to Rochester, she spent a little time and effort on her appearance in order to add yet another side to her multi-faceted personality. Learning to, 'value clothes, to love clothes, to use clothes',[21] seems to have enhanced her reputation and allowed her to make those crucial initial appointments with customers. As impressed on her by her mother's friend and renowned suffragette, Susan B. Anthony, she learned that 'any advertising is good'.[22]

Now a powerful and charming businesswoman, Gleason could be forgiven for taking any credit she could, given the sacrifices she made during her adolescence. However, she stuck to her morals, famously writing to the *New York Times* to correct their description of her as the designer of the bevel gear planar. She modestly added, 'the nearest I have come to designing it is in having a father and brother smart enough to do it'.[23]

Sadly, the strain of their work commitment started to affect the relationship between Gleason and her brothers. They did not agree with her customer-focused approach nor her pricing strategy and this led to tensions, especially after William Gleason took a step back and let three of his four children take over leadership of the business. The respect Gleason had built up throughout her career now irritated James and

Andrew, and her love of luxurious fashion and home design had become an embarrassment. It is possible that Andrew's characteristic bitterness was due to him being the youngest son and last of the children to enter the family business. His discontent came to an aggressive head in 1913, when he announced that he would leave Gleason Works if his sister did not. James took his brother's side, and once again, Gleason was forced to make a sacrifice in order to keep the business running. Despite her compliance with this turn of events, Andrew did not permit her name to be spoken in his house, and his dependants were brought up to have a dim view of the woman to whom the still-operating Gleason Works is indebted. In recent years, a large percentage of overall orders were placed by international customers; how different this may have been if Kate had not had the initiative to promote their products outside of the US.

Gleason carried on despite this sadness, and used her carefully curated skill set to tackle new challenges. For her, 1914 was a busy year as she joined the failing Ingle Machining Company (which she managed to turn around) and was elected as a member of the American Society of Mechanical Engineers (ASME). Gleason had become friends with Lilian Gilbreth (p.94) whilst on a consulting trip with her husband where she discovered Gilbreth learning how to operate a small steam engine whilst sitting in the cab.[24] Gilbreth eventually also joined ASME. Four years later, Gleason took an interest in construction, developing a new concrete-pouring technique that improved the affordability of homes, and still found the time to become president of the First National Bank of East Rochester. She had been unanimously elected by the bank's directors, so took on a position that was extremely uncommon for women to hold at the time. The very drive to put people first that had resulted in her untimely departure from the Gleason Works only got stronger as time went on. After the First World War, Gleason spent time in France helping to rebuild villages that had been affected by collateral damage during the war. Age did not slow her appetite for problem solving; she was mid-project when she died in 1933, aged 67.

To remember Kate Gleason solely as an engineer would not do her justice because she was so much more than that. The beauty of STEM is that it is a tool kit of numeracy, logical thinking, and creativity that can be applied to many challenges; science, engineering, or in Gleason's case, banking, village reconstruction, and architecture. To use Gleason's own wicked sense of humour when telling her brother about a friend who had commented on his similarity to his sister, 'that is one of the most flattering things that can possibly be said about you and you want to remember it'. There's more truth in that than she could have known; it would do no harm if we all had a little more of the Kate Gleason spirit in us.

Further Reading

Marjorie Bell – a British electrical engineer and factory inspector. Standards are a key aspect to quality control, and Bell became the first woman to chair a British Standards Institution technical standards committee.

Caroline Haslett (1895–1957)
Electrical Engineering
United Kingdom

Electricity. For those of us living our comfortable, twenty-first-century lives, it really is difficult to imagine life without it. We may have experienced the occasional power cut, or a weekend of 'off-grid' camping, but even then, we have batteries to fall back on. Born on 17 August 1895, Caroline Haslett grew up in a very different world and she made it her mission to improve access to electricity and make our homes the convenient havens they are today. Haslett's work was often intertwined with the campaign for women's rights, as she used her technical knowledge to provide practical solutions for making the UK a fairer place.

Haslett had a fairly ordinary start to life. She attended school in Haywards Heath, though not much was expected of her due to an [unspecified] ongoing illness that had an impact on her physical strength. Despite this, Haslett managed to complete her education and moved to London to take a business secretarial course so she could find employment in a similar field. By the time Britain entered the First World War in 1914, Haslett had taken her first job at the Cochran Boiler Company. She had initially been offered a role that involved clerical work, however, her job description took a different turn as the reality of war hit the country.

The role of women during the world wars has been highly publicised in recent years. It has become recognised that these conflicts provided women with the option of working in fields that would otherwise be very difficult to access. After the Military Service Act was passed by the UK government in January 1916, many members of the male workforce aged between 18 and 40 left their jobs behind and went to fight for king and country. It would be untruthful to say that women were not able to take part in STEM fields before the First World War, however, the vast majority of those interested in male-dominated subjects had to carefully navigate the realms of social acceptability, or fight the lack of opportunity. Teaching was considered an 'acceptable' female profession, so some women were able to do research alongside their educational role. Even then, there was some pushback from the most conservative individuals; in a report published in 1904 for the Mosely Education Commission, the author expresses concern at the 'ruinous' effect of allowing female teachers to 'contaminate' boys.[1] Education was slightly more progressive in parts of the UK; the University of London had been admitting female students since 1868.[2] Nevertheless, there were many barriers to employment in science, technology, engineering, and maths; even degree holders would take unpaid or low-status roles just to get a foot in the door. Women who were fortunate enough to be wealthy or well connected may have had the opportunity to work with their husbands or family members, often taking on the role of assistant. And then there was scientific writing and illustration: neither of these were considered a real threat to men and, as they were activities that could be done from home and usually focused on topics that were acceptably feminine such as botany or astronomy, there was less resistance. Another difficult argument which gained popularity as scientists started to study anatomy as part of evolution, was that women had smaller brains than men. This was said to affect their intellectual ability, with some going as far as to say that the female sex had evolved to have a mental deficiency. Whether times were changing or whether the government was desperate enough to maintain the war effort, women became eligible to fill vital roles previously occupied by men, which provided an easier route into STEM subjects for many. The government ran many recruitment drives to try and encourage women to take up these roles. Haslett was one of the women who gained opportunities from the economic disruption caused by the war; she was permitted to transfer to the Cochran workshop and undergo engineering training. Part of this training required her to travel to Annan in Scotland. Being so far away from home and learning such a new set of skills would have given Haslett a level of independence and confidence that was highly out of the ordinary given the kind of society she had grown up in. From this time, her ambitions of clerical work were left firmly in the past. She had developed an understanding of the engineering side of the business by the time the war ended, as well as a new appreciation of the kinds of jobs women were capable of. Moving to London had also had an impact on her personal life when she joined the suffragette movement. The post-war years were a hugely important time for campaigners; almost as quickly as women had been encouraged to take up jobs, the inevitable return of many men to the workplace meant that there was no longer the capacity to employ both groups of workers. In the summer of 1919, the UK government passed the Restoration of Pre-War Practices Act, which instructed employers to dismiss female workers in order to make way for returning soldiers. They were under pressure from trade unions to reinstate pre-war working practices although both parties, as well as employers, will have been keen to avoid any unrest that would further damage the post-war economy.[3] The suffragettes also applied pressure, using their efforts during the war as proof that prejudice and discrimination towards women needed to change. The vote was awarded to some women in 1918 although, as she was under 30 and did not own property, Haslett was among many women who did not receive this privilege. Therefore, there were decisions being made that certainly showed progress when compared to the rights of women before the war, but all the changes had amounted to nothing but a taste of freedom, or glimmer of hope for many. There was still much work to be done. Around the same time there was another report written, in which the War Cabinet Committee discussed women in industry. The principle of 'equal pay for equal work' was generally supported;[4] however, finding a consistent method of measuring equal work proved to be challenging and opened up too many avenues for disagreement that in many ways continue to this day.

In 1919, Haslett became secretary for the Women's Engineering Society (WES), as a practical way to advance her feminist cause.[5] The WES was founded the same year that Haslett was appointed to the role, and the same year that the Sex Disqualification (Removal) Act was passed. The law had changed to prohibit disqualification based on gender or marital status for women wanting to join professions or professional bodies. Along with this change in legislation, women who wanted to keep their careers in engineering were in desperate need of practical support, and the work that Haslett started whilst working for WES provided just that. At this time, it was not common for homes to have electricity so heat, light, and appliances were very different to the home comforts we know today. From an early age, Haslett had observed how tiring and time-consuming household chores were.[6] In her eyes, electricity had huge potential to make housework quicker and easier, which would free housewives from drudgery and open up time to do other things. The perfect opportunity arose when she saw a proposal to set up an organisation to encourage the use of appliances and educate the female public about how electricity could be used to provide liberation from endless cleaning and washing duties that shaped their day-to-day lives. Developing countrywide use of electricity was a slow process due to natural fears about new technology, along with the complex infrastructure that would have to be put in place decades before the National Grid was even a concept. In 1924, Haslett co-founded the Electrical Association for Women (EAW); according to Caroline Davidson, author of *A Woman's Work is Never Done*, Haslett was partly responsible for the uptake in home energy usage for housework purposes. Davidson also mentions that the EAW was 'the first female pressure group to have any effect on women's work in the home'.[7] Haslett also had a hand in creating the first consumer information and education body – an incredibly necessary achievement given the sharp rise in consumerism that would go on to define a large part of the twentieth century.

During the Second World War, the Institute of Electrical Engineers formed a committee to discuss the logistics of rolling out electricity in post-war Britain. Haslett was the only woman among her twenty colleagues. One of their main achievements, which has had an effect on our living spaces today, was recommending safety features for plugs and sockets. Her impressive list of firsts continued after the war and in 1947 her contributions were recognised when she was made a Dame Commander of the Order of the British Empire.

Haslett suffered several coronary thromboses and passed away on 4 January 1957 at the age of 61. A year later, the Caroline Haslett Memorial Trust was established; even after her death, the work Haslett did best in life continues, supporting generations of female electrical engineers.

Further Reading

Hertha Aryton – a British scientist and engineer who worked on reducing the noise and flickering of electric arc lighting that lit the streets of London.

Lillian Gilbreth (1878–1972)
Industrial Psychology, Ergonomics, Industrial Engineering
United States

Lillian Gilbreth (née Moller) shaped STEM to such an extent that she has been called, 'The First Lady of Engineering'.[1] And it cannot be denied that she achieved enough to make her worthy of such a title. Gilbreth worked with her husband, Frank, to advance the field of industrial engineering and revolutionise efficiency in the workplace by applying psychology. Their work has become relevant for all businesses to this day as they were interested in reducing waste and improving productivity, both of which led to increased profit. We also have Lillian Gilbreth to thank for her ideas of improving efficiency and personalising the home; changes that promoted a great deal of female emancipation, as modified kitchen layouts and introduction of labour-saving appliances meant that housewives no longer needed to spend so much time looking after the home. Many women will have used this extra time to go to work, or build on their education.

Gilbreth was born in California in May 1878, to a sickly mother and a father who though loving, sadly was not keen for his daughters to have a university education. Because of this, Gilbreth had not taken the classes required to apply to the University of California. Clearly a determined girl, with a little encouragement from her aunt who had studied psychiatry with Sigmund Freud, Gilbreth managed to convince her father that she should give university a go; after all, it was free and she would be commuting from home as was the norm for women at the time. Though her speciality was English, she also studied philosophy and education which allowed her to gain a teaching qualification. She received her bachelor's degree in 1901 and master's degree in 1902, although a period of illness meant that she had to transfer from New York's Columbia University back to her local university in California.

However, it seemed that Gilbreth was destined to be in New York. Shortly after starting a PhD in the psychology of management, she took a break to go travelling in Europe and met her chaperone's cousin before she had even crossed the Atlantic. Frank Gilbreth was of such interest that they were married in 1904, a year after meeting, and moved to New York where Frank's contracting business was located. The couple became equal partners in the firm, where they worked together successfully as well as having twelve children. The 1948 film, *Cheaper by the Dozen*, was written by two of their children and showed how the organisational principles applied by Gilbreth Incorporated were also applied at home.

At work, the couple conducted time and motion studies, building on the work of Frederik Taylor, who used time studies to develop scientific management techniques. Despite being a big player in his field, Taylor was often criticised for his low opinion of workers. His methods aimed to increase process efficiency. However, this was mainly to increase profit for the employer and did not place any emphasis on the importance of well-being and welfare of employees.[2] Naturally, Taylor was not popular with labour unions; he was not shy about voicing his opinions, which could be abrasive. One such opinion documented in a fellow researcher's book, *The Literature of Scientific Management*, stated that time was deliberately wasted by employees so they could lie about how long tasks took to complete.[3] In contrast to this, the Gilbreths created an approach that involved asking both shop floor managers and the relevant labour union to sign their consultancy contracts. Additionally, they tried to ensure that non-financial rewards were difficult for management to take away from their employees, whilst any financial profit that resulted from the increase in process efficiency would trickle down the pay grades and not remain at the top.[4] It was perhaps due to Lillian Gilbreth's background in psychology that she made the (now obvious) connection between productivity and the welfare of workers. The Lever Brothers, owners of a profitable soap-making business in the UK, showed a similar mindset by building the model village of Port Sunlight for their workers to live in. However, Gilbreth was the first person to formally combine the fields of psychology and scientific management within industrial engineering. The husband-and-wife team created films of work processes, usually those which were repetitive, and analysed them to establish unnecessary movement and, therefore, wasted time. Watching the 'before' and 'after' films of processes such as polishing bars of soap, labelling cartons, or filing office records is oddly satisfying. It is clear to see how the workers' jobs were made easier and, in some cases, less likely to cause injury in the long term. In addition to this, they were known to modify work stations for people with physical disabilities. Their scientific approach was another improvement on Taylor's work, which was not always based on quantitative data. The Gilbreths analysed recorded factors such as movements made per brick and the number of bricks laid per hour, both before and after the change was made. This way, they were able to show a tangible improvement. Another example of their work explains how the task of stamping dates on requisitions resulted in 1,900 per hour when one hand was used, 2,300 per hour when two hands were used, and an impressive 3,050 per hour when two hands and a foot pedal was used – that works out at 61 per cent more work being done in the same amount of time.[5] This is just one of the hundreds of changes the couple were contracted to identify and implement; they later branched out into fatigue studies, which pioneered the field of ergonomics. In contrast, Taylor's one-dimensional view of workers is said to have led to the establishment of Human Resources within companies, as scientific management could not always adequately meet the ethical needs of staff.

Unfortunately, despite her education (her degrees were something her husband did not have) and years of relevant experience, Gilbreth was still a victim of gender discrimination. The couple wrote many books and papers; on some, they were sole authors; on others they collaborated. However, from time to time, Frank was advised to remove his wife's name as co-author because they were concerned that their readers would be 'put off'.[6] In addition to this, after Frank unexpectedly died of a heart attack in 1924, several of their company's key clients made the decision not to continue with their contracts. It is difficult to treat this as a coincidence; the number of accounts they lost seems to indicate that they had no confidence in Gilbreth as a leader. They should have had more faith, as she continued to consult and write for nearly five more decades.

On top of this personal change, the US was undergoing a social change that opened up an opportunity for Gilbreth's future career; households that employed full-time servants were becoming less and less common due to an increase in the variety of industry jobs available, and the introduction of appliances such as washing machines and fridges were starting to make the home a more technical environment than it had ever been before.

Perhaps it was partly due to the two reasons above that Gilbreth changed tack slightly, leaving behind the engineering industry that had treated her so ill. Having twelve children to raise and finance would be a daunting task for anyone, and even more so during a time when female breadwinners were uncommon. Instead of increasing efficiency in factories, she applied her logical brain to developing the kitchen as we know it today. She popularised the concept of the 'work triangle', which promoted maximum efficiency by changing the layout of the fridge, sink, and cooker within the kitchen. An exercise she recommended to the public is still surprisingly relevant today. She suggested going about a task in the kitchen, for example, preparing a simple meal whilst trailing a ball of string behind you, so that you can see your movements after you have finished.[7] It's an interesting exercise, and for many of us will be a good indication of how inefficient we can be when not paying attention. Gilbreth did not have a talent for either housework or cooking,[8] so bringing efficiency into the home gave her the benefit of being close to her children, whilst continuing her work on developing household processes to take as little time as possible. There are some parallels to be noted between her work in industry and in the home. Just as the Gilbreths' original work had built on the existing theories of Frederik Taylor, Gilbreth was not the first to consider bringing efficiency into the domestic environment. The fields of home economics and household engineering were already established by the time she was looking to change the direction of her career; Christine Frederick and husband-and-wife duo Martha and Robert Bruère had already published books on efficiency in the home. Frederick in particular was influenced by Taylor's theories, and played on the idea that industrial management techniques could be applied to the role of a housewife (although in reality, the housewife would have to carry out the tasks as well as plan them). Like Gilbreth, Frederick made suggestions of how to increase efficiency whilst carrying out tasks. However, due to her lack of experience in scientific management these were not usually built on any scientific method and she was criticised for being vague.[9] Frederick's career path gradually became more commercial as she positioned herself to give 'feminine insight' to companies that sold appliances. Ironically, despite this seeming like a positive step towards female emancipation, it may have made women even more dependent on their husbands to provide the money that would fund the fashionable commodities they had come to rely on. Once again, Gilbreth's background in psychology and her innate understanding of what drives people would bring a fresh perspective to the complex social shift that was happening in middle-class homes. Gilbreth did not see the use of promoting specific products and thoughtless consumerism, understanding that when it comes to lifestyle and the running of a household, one solution would not suit everyone. She focused on conducting experiments in her home, and in her writings encouraged readers to do the same so that the solution would be tailored to their individual situation. Another key point was understanding that the benefits resulting from greater efficiency in an industrial workplace were not the same as those valued in a domestic environment. However, Gilbreth realised that homemakers, like industrial workers, needed to find their work satisfying in order to improve their chance of happiness. She had the sensitivity to understand that despite her aptitude for statistics and scientific explanation, this was not the method that would best allow her to connect with her audience. Fatigue remained an area of great interest, so much so, that the Society of Industrial Engineers appointed her as the Chair of the Fatigue Elimination Committee in 1926.

Adapting to working from her home environment also made it easier for Gilbreth to return to the consultancy work she and Frank had become so well known for. It was during this period that Gilbreth worked with many large corporations, including heading up market research for Johnson and Johnson as they prepared to launch sanitary products, and improving operational efficiency for floor and admin staff at Macy's.[10] This was no mean feat. Although many corporations had started to realise the value of middle-class women and housewives when it came to targeted sales campaigns, women were not treated with such respect in the workplace and gender discrimination within management structures was commonplace.

Whilst working as a consultant for General Electric, Gilbreth also interviewed over 4,000 women so the design and height of countertops and appliances could be influenced by data. Additionally, many aspects of kitchen design we now take for granted such as fridge shelves, wastewater hoses for washing machines, and bins with foot pedals are also thanks to Gilbreth. Perhaps this makes her the mother of all 'life hacks'. Gilbreth's career progressed even further after her presentation to a gas company led to a collaboration that involved designing efficient kitchens. This in turn gave her the opportunity to work with the New York Herald Tribune Institute, designing a series of kitchens for different family situations. Many old-fashioned kitchens will have been large enough to accommodate several servants but, as times changed, the need for smaller kitchens that accommodated smaller households became necessary. One such radical design even catered for couples who both worked.[11]

In 1935, Gilbreth continued to develop her talent for teaching when she was appointed by Purdue University as a professor of management. As she worked for the School of Mechanical Engineering, this made her the first female engineering professor in the US.

Over time, Gilbreth's professional opinion became increasingly influential, especially so when the public started to realise that a mother of twelve must truly know what she was talking about when it came to domestic issues. Aside from her notable contribution to management theory, perhaps part of what makes Lillian Gilbreth such an important figure in the history of science was her sense of compassion, which probably expressed itself as early as her days of being a young carer for her siblings and mother who suffered from ill health.

As well as looking after a large family and maintaining a career that spanned decades, Gilbreth managed to find time to volunteer: from being a consultant to the Girl Scouts from 1929 onwards, to heading up an unemployment campaign during the Great Depression. During the Second World War when women were required to join the workforce and upskill at an unprecedented rate, Gilbreth advised organisations, the government, and the US navy on labour issues as well as education.

Gilbreth continued to advise, write, and hold a number of speaking and teaching posts until 1968, when her health caused her to stop engaging in public activities. She died in 1972, aged 93, leaving behind a legacy that has inspired several generations, not only in using efficiency to help businesses, organisations, and homes to be the best that they can be but also to seriously consider psychology and the effect processes have on people. Gilbreth paved the way in challenging businesses to think about the human side of their workforce and to adopt responsible management practices. Along with similar work on management theory conducted by Mary Parker Follet, Gilbreth's ideas would go on to be adopted by Japanese managers, who created the system of 5S (Sort, Set in Order, Shine, Standardise, Sustain) that is widely used in manufacturing firms to this day.

Further Reading

Ellen Henrietta Richards – an American safety engineer and environmental chemist who was a proponent of the idea that domestic work was an important part of a successful economy. By applying chemistry to the field of nutrition and other scientific principles to the home environment, she is said to have founded home economics.

Emily Roebling (1843–1903)
Project Engineering
United States

'Under the old common law, a married woman was classed singular with parties incompetent infants, lunatics, spendthrifts, drunkards, outlaws, aliens, slaves and seamen.'[1] This extract from an essay entitled, *A Wife's Disabilities* written by Emily Roebling herself, gives us a glimpse of how her section of society had long been perceived. However, her extraordinary efforts in overseeing the completion of the Great East River Bridge (the Brooklyn Bridge to you and me) must have raised serious questions about the prevailing attitude regarding the incompetence of women, and married ones at that.

Born in the state of New York on 23 September 1843, Roebling may have been fortunate enough to receive an education at the Georgetown Visitation Academy in Washington DC, but she never imagined that she might one day have an intimate knowledge of engineering. Her life changed when she met Washington Roebling at a dance, whilst visiting her brother's army headquarters. Washington's father, John Roebling, was the well-known civil engineer and designer of the Brooklyn Bridge. When the couple married in January 1865, Emily's father-in-law was still at the design stage of the bridge. The groundwork of bridge-building was a great challenge that required watertight chambers to aid workers in digging and laying foundations in the riverbed. As Europe was leading the way in this pioneering use of caissons, the newlyweds dutifully headed across the Atlantic to observe and learn. Whilst in Germany two years later, their first and only child was born.

Just as construction was about to begin in the middle of 1869, the story of the Brooklyn Bridge took a dramatic turn of events. Roebling senior had his foot crushed between the dock and an incoming barge and, although his toes were amputated, he developed tetanus. Within a month, he was dead.[2] His son took the position of chief engineer and wasted no time in starting where his father had left off. But he too was not destined to see the project through. Roebling must have watched in horror as her husband started to develop what we now call decompression sickness, a condition that develops when gases dissolved into the bloodstream at high pressures escape and form bubbles inside the body. Spending so much time in the caissons of the East River had subjected Washington Roebling's body to a dangerous cycle of increasing and decreasing pressure. Over time, he became deaf, blind, mute, and partially paralysed.[3]

First there were three, now only one remained. And she happened to be a woman.

Roebling took what would have been very unusual action, and carried on Washington's work, acting as his eyes and ears on site whilst he directed from his bed. The activity around this project must have been incredible. As well as day-to-day issues on site, there will have been suppliers to manage as well as checking the quality of building supplies, not to mention dealing with the politicians who wanted to be involved. One of the larger scandals reported was to do with the claims of a fraudulent supplier. John Roebling had over-engineered the truss system of the bridge to be six to eight times stronger than he believed it should be, and the contract for the cable was awarded to a company that was able to supply the necessary strength. It transpired that that one batch of cables was sent out and checked by inspectors, only to be switched with cables of inferior quality before the goods were delivered to the construction site.[4] As a result, the strength of the truss was a fraction of its intended value but still four times stronger than it needed to be. The over-estimated strength calculations were fortunate for this reason, but they also future-proofed the structure. Whilst it was designed years before the car was invented, recent figures show that over 100,000 vehicles cross the bridge every day.[5]

Although there is no evidence that Roebling dealt with this supplier situation personally, it is an example of the kind of issues she may have had to oversee. It would be enough to exhaust any modern-day project manager with access to the advanced planning software and communication systems we have in the twenty-first century, but Roebling took it all in her stride, with no small degree of charm and diplomacy. It must have been unimaginably daunting – after all, she had received no technical education and had no engineering qualifications. Willingly taking charge of hundreds of workers, who had probably never had to answer to a woman before, shows Roebling's admirable strength of character.

It's somewhat unsurprising that the day came when the suitability of this arrangement was questioned by those with a stake in the project, and it was suggested that a new chief engineer should be appointed. It was Roebling's speeches that convinced them that between the husband and wife, the work would get done and that there was no need for Washington's title to be given to someone else. And so the couple continued with their mission; it is said that they watched the building site with binoculars from their home and that the hourly reports Roebling provided were so accurate that her husband was able to plan the workload for the contractors very effectively.[6] Washington must have thanked his lucky stars that he had such an able wife; after all, she was the one person who had prevented what could have been the biggest failure of his career.

Brooklyn Bridge was completed in 1883 and officially opened on 23 May. It is 1,834m long, took over thirteen years to build, and claimed an estimated twenty-seven lives during construction.[7] Roebling was the first person to officially cross the bridge and was honoured in a speech at the opening ceremony.

Today, a plaque that features her name can be found on the bridge; a lasting monument to her achievement. After the completion of the bridge, Roebling was as active as ever, earning a law degree at the age of 52, joining and leading several clubs, providing assistance during the Spanish War, and being involved in the New Jersey Board of Lady Managers.

Chapter 7

Healthcare for Children

Anna Freud (1895–1982)
Psychoanalysis
Austria-Hungary, England

Anna Freud was the daughter of the founder of psychoanalysis, Sigmund Freud, and her journey to adulthood was not an easy one. Born in Vienna, Austria in 1895,[1] her family enjoyed a comfortable degree of wealth, but in Anna Freud's case, this privilege did not equate to her having a charmed upbringing. The feeling of competitiveness towards her five older siblings was intense, regarding her eldest sister Sophie in particular,[2] and she did not have a close relationship with her mother. Her complaints of disturbing thoughts[3] may have contributed to bouts of depression and problems with body image[4] which led to her being sent away to health farms. Freud struggled to think about what she should do after leaving school in 1912.

Although she reportedly said that she did not learn much at school, Freud started a teaching apprenticeship in 1914. In the following few years, she earned more responsibility and was offered a four-year contract. During this time, she also joined her father in analysing patients and volunteered at Baumgarten Children's Home looking after Jewish orphans after the end of the war.

The situation looked bleak in 1920 when she was forced to stop working due to illness; however, her aptitude for teaching and interacting with children, along with her interest in her father's work, would help her carve out a niche in the years to come. Although intrigued by the theory of psychoanalysis, it was clear that Freud was looking to find a practical application that would help develop children's emotional well-being. This is understandable as Sigmund Freud was keen to impress that psychological issues and neurotic behaviour exhibited by adults was very likely to have been caused by events that occurred during childhood. 'Early intervention' is a term used in other fields to this day, but Anna

Freud hoped that helping children whilst they were young would enable them to thrive emotionally as they progressed into adulthood.

By 1922, she had become a member of the Vienna Psychoanalytical Society and had presented a paper entitled Beating Fantasies and Daydreams. The following year, Freud started her own practice and not long afterwards, the father-daughter duo started another four-year long stint of analyses. Her personal life changed forever when the children of Dorothy Burlingham were ushered into the Freuds' analysis room. She soon developed a close partnership in both work and life with their mother[5] who was heiress to the Tiffany & Co. New York jewellery retailer. Around the same time, Freud went back to her teaching roots, this time in the specific subject of child analysis. She was fast becoming an influential voice, publishing a book on the topic in 1927 and helping to found the Matchbox School[6] to apply psychoanalytic principles to the educational field. The teaching was project-focused rather than sticking to a rigid curriculum and had the aim of encouraging creativity – the concept of which is still a relevant topic of conversation to those involved in the development of educational techniques today. Although the school was operational for a relatively short time, closing in 1932, it inspired several schools to adopt similar teaching methods. It also laid the foundations of Freud's later work supporting the psychological needs of war-orphans who had experienced trauma.

In the mid-1930s, Freud developed a theory for what is now commonly referred to as 'defence mechanisms'. It is behaviour that the majority of us exhibit on a fairly regular basis. Freud was fascinated with this and published *The Ego and the Mechanisms of the Defence* in 1936. She proposed that defensive behaviour could be broken down into ten distinct types:[7]

- Denial
- Projection
- Turning against the self
- Sublimation
- Regression
- Rationalisation
- Intellectualism
- Reaction formation
- Displacement
- Fantasy

The following year, Freud set up a new institution, the Jackson Nursery, which specialised in caring for toddlers from deprived backgrounds. This endeavour was to be short lived; as Jews living in post-Anschluss Austria, it was a dangerous time for the Freud family. In a successful attempt to find safety away from the anti-Semitic Nazi regime, they fled to London and started a new life.

The emotional, physical, and economic trauma caused by the Second World War affected countless people who lived through it, and Freud and Burlingham had their sights set on helping the youngest in society. They founded the Hampstead War Nurseries to offer foster care for children who had lost parents or their homes during the war. Freud and Burlingham assembled a team of staff, who went on to use observational methods to understand the development of these children and help them to process the horror of what they had lived through. Throughout the six years following the end of the war, Freud also worked with children who had survived Theresienstadt concentration camp and been expatriated to the UK.

In 1952, the Hampstead Child Therapy Course and Clinic, which Freud and Burlingham had opened the year before, was granted charity status,[8] marking the beginning of their legacy. The first generation of child psychotherapists to work for the NHS were trained here.

In the 1960s and 1970s, Freud published *Normality and Pathology in Childhood* and developed links with universities. She became a visiting lecturer at Yale Law School in the US, whilst her clinic encouraged cooperation between academics and paediatric mental health professionals. Throughout her career, she consistently maintained that children are beings in their own right; that they have their own feelings and opinions, and are not simply there to conform to the ideals of adults.[9] This viewpoint was a fundamental step in understanding the child mind, and promoted discussion and action to combat maltreatment and neglect of the most vulnerable people in our world.

The year after Freud died, the Hampstead Child Therapy Centre was renamed in her honour. The decades that followed continued to see advancement, as more rigorous scientific methods were applied to the field that has caused so much controversy for over a century. The Anna Freud National Centre for Children and Families has since teamed up with university research groups in order to apply scientific methods to psychoanalysis[10] and provide evidence to support the importance of improving mental health in children. In terms of social justice, we owe a great deal to this pioneering woman who helped us to understand ourselves, be more tolerant of others, and improve the lives of so many children.

Mamie Phipps Clark (1917–1983)
Social Psychology
United States

Mamie Phipps Clark used psychology as a tool, not only to make a contribution to STEM but to humanity as a whole. Studying and working at a time when racial segregation in schools was the norm in parts of the US, Phipps Clark's work went on to provide important evidence in the Supreme Court's Brown v. Board of Education decision in 1954, which ruled that facilitating segregation in public schools was unconstitutional.[1] A year later, an order was made for states to end segregation in schools 'with all deliberate speed'.[2] Phipps Clark's studies on hundreds of children provides a detailed explanation about just how damaging segregation in schools could be, and the systemic effect it had on the perceptions, confidence, and mental health of the children involved.

As with many great pioneers, Phipps Clark gathered a broad range of experience during early adulthood. Having received a scholarship to Howard University, she enrolled in 1934[3] and studied mathematics and physics. However, Phipps Clark also had an interest in working with children; recognising this, her future husband, Kenneth, encouraged her to read psychology instead. She earned both a bachelor's degree and a master's degree, and entitled her thesis 'The Development of Consciousness of Self in Negro Pre-School Children'.

Building on this research, the Clarks travelled extensively, conducting what they called 'The Doll Study'. The couple was interested in gauging racial perceptions of young children, based on a simple experiment that involved presenting them with four dolls. Each doll was identical in all but skin colour. As they asked the following questions[4] over and over again, they started to determine just how much the perception of colour and race had on the identity and self-esteem of children.

> 'Show me the doll that you like the best or that you'd like to play with.'

'Show me the doll that is the "nice" doll.'

'Show me the doll that looks "bad".'

'Give me the doll that looks like you.'

Kenneth Clark is reported to have said that when asked the last of the questions above, some children experienced a highly emotional reaction and would 'cry and run out of the room'. Whilst it might be expected for white children to identify with, and associate positive attributes to, the white doll, disturbingly, it also came to light that some black children also shared this view. This starts to give us some insight into the dark truth about how children can and have been influenced to believe dominant societal norms. The experiments conducted by the Clarks built and expanded upon the work of husband-and-wife duo Ruth and Eugine Horowitz, who had used a smaller sample group of nursery school children to explore self-identification in terms of race.[5] Both couples recorded similar findings, in that many of the black children they spoke to incorrectly identified themselves as white.[6] Similar findings were later recorded by Alyce Gullattee (p.41). In her article 'The Negro psyche: fact, fiction and fantasy', she presents case studies that include a young black child wearing a sunhat when playing outside to protect her perceived white skin and blonde hair.[7]

Even after receiving a doctorate in experimental psychology from Columbia University, Phipps Clark struggled to find a job. Her PhD supervisor had discriminatory views on the intellectual difference between black and white people, so he offered no encouragement.[8] The first two jobs Phipps Clark tried did not suit her, as her high level of education made her overqualified. Eventually, she was successful in gaining a job as a psychologist at the Riverdale Home for Children in New York, which provided accommodation for homeless black girls. As she started to interact with her young patients, Phipps Clark started to understand the extent of the lack of services available to black and minority children. Her observations during the unsuccessful attempts to improve social services in Harlem led to her taking action. In 1946, Phipps Clark used a loan from her father to open what is now known as the Northside Centre, to help with child development and provide basic support for families. She served as director until she retired in 1979; services were added over the years, including therapy, career guidance for teenagers, training for black parents, and tutoring.

It is an absolute testament to the talents and character of Phipps Clark that she worked so hard to back up the fight against racial injustice with science, despite living in a time when she may not have been permitted to use the same facilities as white people. The Supreme Court finally declared all forms of segregation unconstitutional in 1968. As struggles against racial discrimination continue, Mamie Phipps Clark remains a shining beacon of inspiration to anyone who shares her passion for using science as a power for positive social change.

Virginia Apgar (1909–1974)
Medicine, Anaesthesiology, Obstetrics
United States

Virginia Apgar dedicated her life to developing what was then the new field of anaesthesiology, and made important observations and improvements to how the health of babies is assessed in the minutes following delivery. She was born in 1909 in New Jersey; comparing the infant mortality figures from that year with 2020 show the extent of the advancement of obstetrics in the US, as well as the rest of the world. In the year Apgar was born, 194 0–5-year-olds died per 1,000 born, with this number dropping to just over 6.2 in 2020.[1] Globally, the infant mortality trend has also decreased over time in every country.[2]

Apgar was keen to pursue medicine at an early age, and studied zoology for her undergraduate degree, graduating in 1929 from Mount Holyoke College before completing medical school in 1933.[3] After this, she became a physician specialising in anaesthesiology and worked at Columbia University, eventually becoming the first female professor at the College of Physicians and Surgeons in 1949.[4] Shortly after completing her own residencies, she was not only coordinating them but training both students and doctors in what was a relatively new area of medicine as well. On top of administration and developing education programmes, Apgar was involved in research. She had a particular interest in lowering mortality rates of newborn babies, as well as how anaesthetics administered to mothers affect the child.[5]

Apgar used her knowledge of anaesthesiology and specialist interest in obstetrics to create the 'Apgar Scoring System'. This uses five important signs to assess the health of newborns and determine whether they need medical attention.[6] In her 1953 publication, 'A Proposal for a New Method of Evaluation of the Newborn Infant', she notes how there were several existing methods of establishing health, such as time taken for the first breath or cry. However, with anaesthetics added to the picture, relying solely on these observations could lead to the wrong conclusion.[7] Apgar's system was simple yet clear. It was designed to cause minimal interference after delivery; to prevent significantly different scores being given by different members of staff, the signs chosen were not overly subjective.[8] These five signs – heart rate, respiration, reflex response, muscle tone, and colour – were each given a rating of 0 to 2. By testing neonatal blood, she and her colleagues were able to show the efficacy of using this testing system minutes after birth. As such, it soon became standard practice. Having a standardised scoring system allowed the health of the newborn to be linked to delivery methods and use of anaesthetics, among other factors, which are different for each birth.[9]

Apgar became interested in the detection of birth defects and took on a range of pivotal roles for the March of Dimes Foundation throughout her career. In 1972, she published a book with Joan Beck entitled, *Is My Baby All Right?* Having such a passion for her work, Apgar never retired and worked until her death on 7 August 1974, aged 65.[10] By this point, she had also held teaching positions at Cornell University School of Medicine and Johns Hopkins School of Public Health, advancing education surrounding teratology (the study of birth defects). Using her scientific knowledge and compassion in equal measure to provide clarity for humanity at the times when we are at our most vulnerable, Apgar truly changed the world for newborns and their families.

Helen Taussig (1898–1986)
Paediatric Cardiology
United States

Another inspirational woman in STEM who made valuable contributions to improve child mortality was Helen Taussig. Born on 24 May 1898, like Apgar, Taussig grew up in a time when the health of neither children nor adults in America could be taken for granted. Her mother, Edith Thomas Guild had studied biology and zoology at Radcliffe College, making her one of the institution's first female graduates, but it was a tragedy that she contracted tuberculosis and died when Taussig was just 11 years old;[1] she would never see the impact her daughter had on the world.

Taussig's school years were challenging. Not only had she lost her mother but she herself contracted mediastinal tuberculosis, which left her unable to attend full days at school for a number of years.[2] Additionally, whooping cough had left her partially deaf; her hearing had completely disappeared by the time she reached adulthood. Taussig also had dyslexia, but thanks to her father's gentle persistence and encouragement she was able to get through grammar school and have her application to Radcliffe College accepted like her mother. Frank William Taussig was a professor of economics at Harvard for over four decades and played a key role in developing the theory of international trade through his writings. He was seen as such an influential economist that in 1919 he was invited to Paris by President Woodrow Wilson to advise on commercial treaties. Thanks to his devotion to his family as well as his career, Frank Taussig's daughter was also able to find her passion. Despite having struggled with reading, she earned not only a bachelor's degree but also went on to study medicine at Johns Hopkins University when she realised she would not be permitted to receive a qualification at Harvard (her first choice) even if she attended classes. She earned her medical degree in 1927,[3] and decided to specialise in paediatrics, combining the field with her growing interest in cardiology.

Being deaf did not seem to hinder Taussig in her career, she even learned to use her sense of touch to monitor patients' heartbeats. Her real success began when she started observing babies who had a blue tinge to their skin shortly after birth. This was commonly due to a heart defect called Tetralogy of Fallot. The aorta and pulmonary artery of a foetus are connected by a blood vessel called the ductus arteriosus. This vessel usually closes a few weeks after birth; when this does not happen, infants must undergo surgery. By contrast, Taussig had noticed that children with conditions such as Tetralogy of Fallot were more likely to survive and suffer fewer symptoms if the ductus arteriosus remained open. Her theory went against standard medical practice of the time so she had considerable difficulty finding a surgeon willing to try it out. Two years later, Johns Hopkins had a new chief of surgery, Alfred Blalock, and Taussig was able to convince both him and his laboratory assistant, Vivien Thomas, to develop the surgery with her. The Blalock-Thomas-Taussig

Shunt was a resounding success and is still used to this day to prolong the life of babies whilst they wait for more complex surgeries to be attempted.[4]

An advocate for unborn and newborn babies her entire life, Taussig actively campaigned to block the regulatory approval of thalidomide in the US. Due to her efforts to educate fellow healthcare professionals and politicians on the dangerous effects of the anti-morning sickness drug, it was banned in the US as well as Europe.[5]

Chapter 8

Understanding Our World

Inge Lehmann (1888–1993)
Seismology
Denmark

It can often seem like the only way to have an established scientific career is to do well at school, get into university, complete summer placements, achieve a top grade then have years of meaningful experience. The story of Inge Lehmann's life offers great hope that this does not necessarily have to be the case.

Lehmann certainly started off life as you would expect a scientist to. Born on 13 May 1888 to a psychology professor and his wife, Lehmann was sent to a private school in Copenhagen, run by Hannah Addler, the aunt of physicist Niels Bohr. This school was unusually progressive for its time as it taught both boys and girls; as such, it appears to have instilled in the young Lehmann the idea that gender has no impact on intellect. She later stated that this led to a rude awakening during her career as she encountered people who did not have the same attitude.[1] Although it is not possible to know if, or to what extent, she was discriminated against in the workplace, the campaign for women's suffrage was taking place throughout most of her twenties, showing the social climate she was living and working in.

In 1907, Lehmann went to the University of Copenhagen to study mathematics.[2] During this time, she also took physics, chemistry, and astronomy classes, and in 1910 moved to Cambridge to continue her studies.[3] However, by the following year she was suffering from exhaustion and fell ill. Forced to move back to Denmark, Lehmann worked as an assistant to an actuary. Although the job she was doing (determining the risk that is used to calculate insurance rates[4]) was far from her interest in science, it proved to be valuable experience which developed skills that would be beneficial later in her career. This is an encouraging and important lesson relevant to anybody pursuing a career in STEM. Lehmann didn't return to university and graduate until 1920 and even after that, she spent three years of graduate life as assistant to a professor of actuarial science.

However, her patience and ability to convert an unplanned situation into opportunity was about to pay off. In 1925, Lehmann was offered a job as assistant to Professor Niels Erik Nørlund, director of the institution Den Danske Gradmaaling.[5] Nørlund was an expert in seismology, a subject Lehmann had first become interested in as a teenager when she experienced an earthquake first-hand in her Copenhagen home. Lehmann's important role was to set up seismological stations in Denmark and Greenland, along with three young male colleagues. The seismograph, an instrument that records seismic waves, was invented in 1880[6] and it was this that Lehmann and her colleagues had to understand in order to carry out their work. As none of them had ever used this machine before, Lehmann spent these years learning from books and studying with Beno Gutenberg, an expert in the field. Gutenberg would go on to mentor the man who invented the scale for measuring earthquake intensity, Charles Richter.[7] By visiting seismic stations in Germany, the Netherlands, and France, Lehmann quickly gained a large amount of knowledge about the measuring of seismic activity and how it could reveal the composition of the earth's interior. By the end of 1926, the first observatory in Copenhagen was in action and Nørlund's instruction to only use the highest quality seismographs meant it set a gold standard for future installations. This was confirmed when Lehmann later conducted investigations into the accuracy of European seismic stations and found that Copenhagen's was within the top five.

After attending a conference on geodesy and geophysics in Prague in 1927, Lehmann started an ongoing correspondence with British seismologist Harold Jeffreys and her mentor Gutenberg. The conversation shared between the three scientists would go on to have a lasting effect; Lehmann's future paper was built on the findings shared by Jeffreys and Gutenberg, and they in turn would confirm the accuracy of her future work.

The reward for Lehmann's patience and the turning point in her long journey to carve out a career presented itself in 1928, when she completed her master's degree in geodesy. Shortly after this, Nørlund appointed her as State Geodesist and head of the Department of Seismology at the Geodetical Institute of Denmark. This was fortunate timing as shortly after her appointment, two earthquakes took place that would have a great impact on Lehmann's understanding of what lies below the earth's crust.

Lehmann was title author of The Earthquake of 22 III 1928. In this work, she explains that although the epicentre was in Mexico, readings were recorded as far away as Europe. In a great effort to achieve international collaboration, seismological stations from Mexico, the US, and Europe sent data to the Geodetic Institute in Copenhagen. She mentions that the American stations are positioned between 30 and 45° from the approximate epicentre and how 'at distances greater than 30°, the residues are small'.[8] In summary, this fourteen-page write up discusses the method she used of looking at and comparing data from these different stations to calculate an accurate time-curve. This is a graph of the time taken for a seismic wave to reach the station from the epicentre. This meticulous compilation of data was a step forward in improving the accuracy of earthquake measurement and recording.

The Murchison earthquake that happened in New Zealand the following year offered another opportunity for learning. From 1800 onwards, it was thought that the earth's liquid core was surrounded by a solid mantle, which was in turn surrounded by a solid crust. 'Discontinuities' was the name they gave to the areas of different densities that separated these layers. There was also an understanding that in the aftermath of an earthquake, two types of seismic wave could be detected: compressional P-waves (pressure waves that travel downwards) and transverse S-waves (shear waves that travel sideways). The molten core theory seemed sound as transverse waves are unable to penetrate liquid. However, the P-waves detected from the Murchison earthquake raised questions as they seemed to turn up, albeit faintly, on seismographs that shouldn't have detected them – given the physics of earthquake wave behaviour based on existing models. Again, Lehmann encouraged participation of the international community. She pored over the data, and where others may not have had the patience to look into the very infrequent occurrence of P-wave detection, Lehmann took the time. Slowly but surely, she began to build a controversial argument that the core of the earth is, at least in part, solid. The weakness of the detected P-waves could be explained by them entering the outer liquid core but then bouncing off this solid inner core.

By 1936, she had gathered enough evidence to publish her hypothesis in her paper, called 'P'.[9] The area in which wave velocity increases became known as the Lehmann Discontinuity. Backed up by Gutenberg and Jeffreys, Lehmann's findings became widely accepted and were eventually confirmed by computers in the 1970s.

Lehmann was one of the longest-living female scientists, passing away at the grand old age of 104.[10] Although she chose a career over marriage and children, a decision that seems unacceptable from a privileged twenty-first-century perspective, family and friends alike have commented on her warm character both within and outside of the workplace. Her nephew fondly recalls her creating a filing system for international seismic data out of cardboard oatmeal boxes,[11] and in the January 1994 volume of *Physics Today*, Bruce Bolt called Lehmann a 'remarkable woman', stating that she was a 'strong and independent person' who was full of energy and personal warmth.[12] This charming part of her personality has been noted by others. Jeffreys' wife remembered that after seeing her hostess using a tea set that 'had a pattern imitating Royal Copenhagen', Lehmann had gifted them the 'real thing' during her next visit.[13]

There is much to learn from Lehmann's character and how her patience and perseverance was critical to her success. Where others may have given up and taken an easier route, she threw herself into work that was not necessarily relevant to her interests. Through taking on such work, she was able to learn transferrable skills that would prove to be invaluable and give her a head-start later on in life.

Maria Goeppert Mayer (1906–1972)
Theoretical Physics
Germany, United States

For the woman who discovered the structure of the nucleus, Maria Goeppert Mayer spent much of her scientific career carrying out unpaid work. She countered this with her intelligence and enthusiasm to learn, and by building friendships with peers who were able to see her talent for what it was.

Goeppert Mayer had a privileged education. She was born on 28 June 1906,[1] and lived in the university town of Göttingen from an early age. Keen for his daughter not to spend her whole life as a housewife with no interests, her father paid for her to attend private school as this was the only local institution that provided girls with the option of going to university. The school closed down during her time there,[2] but Goeppert Mayer was a budding mathematician by that point and sat the university entrance exam to see if she could continue studying her favourite subject. Her application was accepted, though she was soon inspired by physicist and family friend, Max Born, to study the atomic and subatomic world of quantum physics instead. Her father died in 1927 and her mother started taking in lodgers to support the family. A certain young American chemist who came to stay would capture the attention of Goeppert Mayer almost as much as science had. Fortunately, Joseph Edward Mayer was not looking for a traditional wife and encouraged her to finish her PhD when she appeared to lose interest in her dissertation. Even after they married, moved to the US, and Goeppert Mayer found that her lack of housekeeping skills might make life tricky, her husband insisted that the pursuit of science was more important and talked of hiring a maid. The strict anti-nepotism laws of the time, which were partially a consequence of the economic depression, meant that the couple could not both work at Johns Hopkins University, where Joseph had been offered a job. However, they did recognise Goeppert Mayer's knowledge in the field of quantum mechanics by allowing her access to facilities and projects. After giving birth to their first child in 1933, she continued with her work at the university as a volunteer research associate and started combining her knowledge of quantum physics with chemistry. By crossing scientific fields, she was also able to work with her husband and his colleagues which broadened her knowledge base and would prove useful later in her career. They had such a shared passion that their children would later ban scientific discussions from the dinner table![3]

By the time their second child was born, Joseph had been offered a tenured position at Columbia University. Again, the physics department would not employ his wife. However, the chairman of the chemistry department, Harold Urey, decided to give Goeppert Mayer a chance by offering her a position as an unpaid lecturer.

Back on her home continent, the political climate was changing. In 1933, Adolf Hitler had been appointed as the Chancellor of Germany and the fascist regime in Italy had caused many people to flee the country as refugees. The physicist and Nobel laureate Enrico Fermi was one such refugee and his arrival at Columbia University provided Goeppert Meyer with another source of professional encouragement and friendship. At this time, it had already been established that electrons reside in shells around the nucleus of an atom. At Fermi's suggestion, she set about predicting the structure of the valence (outermost) shell of elements with an atomic number greater than 96 (uranium).[4] This was particularly impressive as these transuranium elements had not actually been discovered at the time; from her work, Goeppert Mayer correctly concluded that the elements beyond uranium would form a whole separate series in the periodic table.

Goeppert Mayer clearly felt that she had more to give and in 1941 became a mathematics substitute teacher at Sarah Lawrence College.[5] Her background in both physics and chemistry provided excellent inspiration to her students, who were more interested in how mathematics could be used in real-world applications than learning the subject for the sake of it. Unbeknownst to the students or teachers at the time, many of the students would soon be using their scientific knowledge to support the war effort and the interdisciplinary education they received was invaluable. By the next year, Goeppert Mayer was teaching four different classes: differential integral calculus, physics, and physical chemistry, on top of her mathematics class.[6] To add to the intrigue, her lessons on radioactivity, nuclear physics, and the conversion of one element to another, were subtly mirroring her other more secret job.

Many scientific projects were initiated at the onset of the Second World War. Because of this, Joseph frequently worked away from home at a US Army facility and childcare responsibilities fell to Goeppert Mayer. Rather than becoming a housewife, she continued teaching and at Harold Urey's request took on her first paid research role. This investigative work, which involved separating an isotope of uranium that can undergo a fission reaction when bombarded with a neutron, was highly confidential because of the implications of a nuclear reaction that releases a huge amount of energy. Goeppert Mayer's research became so time-consuming that she was required to take a leave of absence from Sarah Lawrence College.

In the years that followed she became even more involved in war-related projects, investigating the transfer of radiation through a range of materials as part of The Manhattan Project. The end goal was never disclosed, but the temperatures she and her colleagues were simulating left little doubt that they were working on a project involving nuclear weapons. Personally, this period of Goeppert Mayer's life must have been difficult; not only was her family fragmented with her husband often away from home and her children left with a nanny, but there were also inevitable concerns about her family and friends in Germany and the way in which any potential nuclear weapon might be used. Goeppert Mayer did not achieve any significant results and expressed relief that this was the case.[7] Years before the atomic bomb was dropped on Hiroshima, she had held a seminar about the work of scientists contributing to the development of weapons. It is clear that she was acutely aware that at this point in history, theoretical physics and related fields were anything but abstract science and had the potential to cause great harm. They were uncertain, difficult days; various pressures had caused her to smoke and drink more[8] and a year before the end of the war, Goeppert Mayer became ill to the point where it affected her work. By the time she recovered in the spring of 1945, she returned her focus to teaching at Sarah Lawrence College, deciding to turn down Urey's request for her to work more than half a day every week at the laboratory at Columbia University.

After the war, life changed for the Mayer family as both Joseph and Maria were offered positions at the University of Chicago. Although she was once again classed as an unpaid 'volunteer' professor, they were both able to work at the university's Institute for Nuclear Studies alongside many of their old friends who, like Goeppert Mayer, had had some involvement in the development of the atomic bomb. One of Goeppert Mayer's students from her early years in the US was now the director of the newly established Argonne National Laboratory. Goeppert Mayer was offered a paid part-time research position. Her work became exploratory and revolved around the structure of the atom.

Different chemical elements are characterised by their atomic number, which is the number of protons in the nucleus. Neutrons are the other subatomic particle in the nucleus; when this number changes, an isotope of that element is created. From experimental data, it had been observed that elements with 2, 8, 20, 28, 50, 82, and 126 protons or neutrons (nucleons) were more stable and less likely to decay, as well as being more abundant. It seems that there was not a lot of interest in this 'coincidence' at the time. They were amusingly called 'magic numbers' and Goeppert Mayer was unique in wanting to investigate further. Like the 'layers of an onion', these two subatomic particles are held in different energy levels (shells) within the nucleus. This can be shown because when a shell is full, more energy is required to displace a nucleon, and a graph showing the energy required to displace nucleons from all known elements shows that peaks occur for certain ones. Having a theoretical mindset, Goeppert Mayer set about trying to find a mathematical explanation as to why this was the case and, ultimately, why the magic numbers were significant. One day during a conversation with Fermi, he mentioned the concept of orbital spin coupling. Most of us could be forgiven for thinking that this is not particularly exciting, but for Goeppert Mayer it was the instruction that led to the pinnacle of her life's work. She proposed that nucleons spin around an axis as well as orbiting the centre of the nucleus, in the same way the earth spins and moves around the sun. Her model showed that the energy of the nucleon changes depending on the interaction between the spinning motion and orbiting motion; when aligned there is an increase in energy and when in opposition there is a decrease in energy. The start and end of each shell can be determined by the presence of the magic numbers, which occur when there is a large difference in energy levels.

Around the same time, nuclear physicist Hans Jensen and his colleagues had independently come to the same conclusion in Germany. The models from both countries were published around the same time and Goeppert Mayer started a collaboration that would eventually yield their co-authored book, *Elementary Theory of Nuclear Shell Structure*, in 1955. This understanding of atomic structure played a key role in future scientific advancement.

At the age of 58,[9] Goeppert Mayer finally became a full professor when she and her husband were offered positions at the University of California in 1960. By the time she was awarded her share of the Nobel Prize alongside Hans Jensen and Eugene Wigner for their work on the structure of the nucleus, she had had a stroke and suffered from ill health. Goeppert Mayer died in 1972, having inspired countless students and changed perceptions in conservative scientists she had come across over the decades. She was a sociable woman and used her charm and intelligence to push the boundaries of what society expected of a married woman.

Marie and Irène Curie (1867–1934) and (1897–1956)
Physics, Chemistry
France

One may assume that the first woman to receive a Nobel Prize, the only person to receive a Nobel Prize in two different scientific disciplines, and the first female professor and head of laboratory at the prestigious Sorbonne in Paris, came from a wealthy, successful, and well-connected family. The truth is, Marie Curie (née Skłodowska) was raised in a Polish family who lost their property and fortune because of their involvement in the movement for Polish independence. Although both parents had worked in education, her father lost his job due to his patriotic sentiment and her mother had resigned in order to raise the children. Unfortunately, by the time Curie was 10 she had lost both her mother and eldest sister in the space of three years.

After leaving school, Curie spent two years away from home due to poor mental health[1] and presumably to ponder her future, as her gender prohibited her from applying to institutions for higher education. During this time, Curie also experienced a painful failed relationship: her suitor's family refused to accept their son's choice of wife as she was too poor. She worked as a governess for several years to fund the medical studies of her sister Bronisława at the progressive 'Flying University' which did admit women. This was on the understanding that Bronisława would return the financial favour when it was Curie's turn to study. She continued to tutor whilst furthering her own education until she moved to Paris in 1891. This all paints a picture of a very strong and independent character, and this is true. Curie got to the position of receiving a Nobel Prize by her own merit, not simply because she married a well-established scientist. She did not take the option of marriage to make her life easier – in fact, she did not readily accept Pierre Curie's offer as she was hoping to return to Poland. However, being rejected for a teaching position by Kraków University on account of her being female[2] may have caused her to consider Pierre's offer more seriously. Besides, he had already shown sincere commitment by offering to move to Poland, even if he had to become a French teacher.[3] A dramatic gesture indeed for a scientist who in the future had to be encouraged by his wife to publish much of his research. And so, Curie returned to Paris after spending the summer with her family, married Pierre, and started her PhD and the research that would earn her a place in the history books.

Until this point, Pierre Curie's focus had been on the behaviour of magnetic materials. He had spent years researching this with his brother and during this period identified the Curie point, the temperature above which a material stops being magnetic. However, when Marie Curie became interested in the work of fellow scientist Henry Becquerel, who had observed that uranium salts gave off a strange glow, Pierre decided to join his wife in her research. After grinding down and analysing a range of similar materials, she realised that the energy being detected by Pierre's electrometer was coming from the uranium atoms themselves. She called this effect 'radioactivity' and the name stuck. Whilst working with similar compounds, the couple discovered two new elements: polonium, named after Curie's homeland, and radium. Shortly before this, they had welcomed their baby daughter Irène into the world so it was a busy time. Pierre was nominated for a Nobel Prize for the discovery of radiation in 1903 and at his insistence, Marie Curie's name was also added. However, the fact that she had actually been nominated the year before,[4] shows that her work was recognised and not merely an extension of her husband's.

Tragically, Pierre Curie was killed in a horse and carriage accident three years later, when their second daughter was just 2 years old. Finding herself in a situation that would cripple someone with even the greatest mental fortitude, Curie started to build up her solo career. Her husband had been a professor at the Sorbonne and Curie inherited the chair upon his death.[5] If those in scientific circles weren't already aware of her exceptional work, they certainly were by 1911 when she was awarded a second Nobel Prize, this time in the field of chemistry. She was honoured for her discovery of polonium and radium and the work she had done to determine the properties of the latter. Despite the prestige and recognition that these awards brought her, they were not what kept Curie focused on her career. In fact, both husband and wife had been so invested in their research that they had not even attended the ceremony for their joint Nobel Prize. Curie's attempt to donate her Nobel medals to the French war effort,[6] despite the xenophobia and false accusations of being Jewish that she often faced, shows that she did science for the love of her field and nothing else. She also showed patriotism for her adopted country in the work that she did during the war, developing and designing trucks that held X-ray units. By this point, Curie had understood the potential radiation could have within the field of medicine, so with the help of her eldest daughter Irène, she was responsible for the installation of these trucks that were so invaluable to battlefield surgeons.

It would be fair to say that Marie Curie has become the most famous female scientist in history. She was fortunate that like her daughter Irène and Gerty Cori, who was awarded the 1947 Nobel Prize in Physiology or Medicine, she had a husband who was fiercely unapologetic about defending her work and right to be recognised as a scientist. Admirably, rather than simply accepting certain privileges she enjoyed during her career, Curie set an example of going above and beyond to leave a legacy that has inspired generations, starting with her firstborn.

Irène Curie had a privileged upbringing in terms of education; Marie spent the vast majority of motherhood as a single parent, and decided to personally teach her daughters physics. Lessons in other sciences, humanities, sports, and arts, were left to her colleagues. However, Pierre's premature death and the outbreak of war in the year Irène was set to attend university must have taken its toll. After her involvement with her mother's wartime mission and spending her late teenage years on the front line of battle, she was perceived as cold – mainly by other scientists who resented her privileged family connections. Fortunately, a man called Frédéric Joliot who had been assigned to her at the Radium Institute was able to see past this hard exterior. Curie married him in the autumn of 1926 and, like her parents, entered into a relationship that was simultaneously personal and professional. Despite the brevity of their working partnership, it was extremely productive. It was Marie Curie who first suggested the use of radiation in cancer treatment, but it was her daughter and son-in-law who conducted the pioneering research that led to it becoming a common treatment in oncology today. They used polonium as a radiation source and by bombarding aluminium, discovered artificial radioactivity. For this innovative work they were awarded the Nobel Prize in Chemistry in 1935 which was, happily, announced before her mother's death. In contrast to her parents' relationship, at the beginning of their collaboration Irène was the main authority whilst her husband had to work hard to show that their scientific partnership was made up of equal talent on both sides. She held a strong belief that women were perfectly capable of doing the same scientific work as men, and mentioned that she would like her joint award to go some way to proving this to other French women.[7]

Having a privileged family background did not mean that Irène Curie's life was without trouble. As her own daughter and physicist, Hélène Langevin-Joliot, reminisces during a fascinating interview with the American Institute of Physics, Curie was forced to flee the country with her children during the German occupation of France, leaving her husband behind so he could continue with his secret work supporting the war effort.[8] She would also have observed the dropping of the atomic bomb on Hiroshima, and was very involved in the talks surrounding the future of atomic weapons. As communists, the war years would have been dangerous for the family. Sadly, the other immediate danger was the couple's prolonged exposure to radioactive materials. Curie died of leukaemia and her husband developed incurable liver disease induced by radiation a couple of years later. Although not as well known as Marie Curie, Irène also left an incredible legacy in her children, both respected scientists, allowing the 'Curie Dynasty' to continue for over one hundred years.

Gerty Cori (1896–1957)
Biochemistry
Austro-Hungary, United States

Gerty Cori (née Radnitz) was amongst the first women to be co-awarded a Nobel Prize. After meeting her husband, Carl Cori, at the University of Prague, the couple formed a strong partnership both professionally and personally. Their decades of research led to the vital understanding of how carbohydrates provide the energy needed by the human body to function properly when oxygen is in low supply. In these conditions, lactic acid is produced in the muscles as a by-product of carbohydrate metabolism. This substance reduces the pH and needs to be removed before it affects muscle cells. The Coris discovered that once complex carbohydrates are broken down into simpler sugars that provide energy to muscle cells, this harmful lactic acid is transported to the liver where it is converted back into glucose. This glucose is then transported back to the muscle cells and the cycle starts again.

Despite having worked with her husband from the beginning of their acquaintance, Cori faced many barriers due to her being female.

The Coris emigrated to the US to distance themselves from the anti-Semitic feeling that was starting to develop in 1920s Europe, only to find that many universities exercised strict anti-nepotism rules. The couple worked together for nine years at the Roswell Cancer Institute in New York; however, when trying to move on after this period, Carl was forced to turn down many employment opportunities as the institutions would not entertain the thought of offering a position to his wife. One particular institution went as far to say that Cori should be aware that she was hindering her husband's career by attempting to work with him.[1] The Coris provide a unique example in that although Cori had the same education and knowledge as Carl, it was never her being offered the job and her husband being rejected on grounds of anti-nepotism. Thankfully, Washington University in St Louis, Missouri noticed the couple's potential and saw fit to employ them both; Cori started as a research associate and became a professor after fifteen years, and Carl's initial role of professor of pharmacology developed into chair of the biochemistry department.[2] The Coris' research thrived and they started to study enzymes: molecules that act as catalysts for chemical reactions in cells. Having already developed the theory of the carbohydrate (now Cori) cycle, they went on to identify the enzymes that aid the breakdown of glycogen, the storage form of glucose. As well as this, they synthesised glycogen in a test tube, which helped them understand how it is built up. For the identification of these processes, Gerty and Carl Cori won the Nobel Prize in Physiology or Medicine in 1947.[3]

As well as carrying out ground-breaking experiments, they welcomed scientists to work in their laboratory without discrimination, including women, Jewish people, and refugees. Six of their colleagues would go on to receive Nobel prizes; presumably, the Coris' excellent mentorship and encouragement had some part to play in this. One of their students reminisced that Cori always had time to listen to any problems he had,[4] which shows her to be a great teacher as well as scientist, a quality essential for raising up future generations.

Further Reading

Elizabeth Blackburn – an Australian molecular biologist who received a share of the 2009 Nobel Prize in Physiology or Medicine for her work on telomeres and the enzyme telomerase. Telomeres are regions at the end of DNA that protect the strands and have a role in cellular ageing.

Dorothy Hodgkin – a British biological chemist who was awarded the 1964 Nobel Prize in Chemistry for determining the structure of molecules by using X-ray crystallography. Her major accomplishments include discovering the structure of Vitamin B12, penicillin, and insulin, which made it easier for medication to be developed.

Rita Levi Montalcini – an Italian neurobiologist who discovered nerve growth factor, a substance that promotes growth of the nervous system.

Annie Easley (1933–2011)
Computer Science, Rocket Science, Mathematics
United States
Annie Easley was a pioneering computer engineer whose thirty-four-year career at the National Advisory Committee for Aeronautics (NACA) – which was later rebranded as NASA – saw the early use of people as 'computers', the gradual uptake of the machines we are so familiar with nowadays, and even the start of women starting to wear trousers in the workplace! As her employers started to appreciate her logical mind and ability to learn quickly, her role grew more specialised over time to the point where she had a hand in developing the computer software for Centaur. This rocket used pioneering fuel technology and would eventually be used to power the first American space probe that landed on the moon.

Easley had an extremely supportive mother, who was so encouraging about achieving goals regardless of ethnicity, background, or colour, that even Easley's friends saw her as a role model.[1] Although gifted at maths throughout her time in school, Easley mentioned in an interview with NASA that she felt drawn towards pharmacy as a career because she liked helping people. For two years she studied at Xavier University in New Orleans but left before graduating. She married in 1954 and moved to Cleveland so the couple could be close to her husband's family. Unfortunately, the local college did not provide a pharmacy course so Easley was not able to finish her studies. She worked as a teacher, until one day she saw a piece in the newspaper, detailing the work of two sisters at NACA. With her interest in maths, Easley considered that it might be an interesting opportunity. Two weeks later, her application was accepted and she found herself with security clearance to enter her new office – a building called 'Materials and Stresses'.[2] Easley and the people she worked with were called 'computers' because they carried out the mathematical calculations required by engineers and scientists who were working on problems in the laboratories and test houses. Once they received the calculations, Easley and her colleagues would type numbers into large desktop machines that were the calculators of their time, looking up more complex equations such as square roots and exponentials in tables provided and noting the answers on paper.

However, Easley was destined for greater things than simply doing calculations for scientists. When the term 'computer' started being applied to machines instead of people, Easley learned the programming language FORTRAN, short for Formula Translation, and assisted with a variety of projects: from sending jet planes to measure the damage caused to the ozone layer, to working on shuttle launches. In a period of heightened tensions between the US and the USSR during the Cold War, the first man-made satellite, Sputnik, was fired into space by the Soviets, catalysing the beginning of the space race. Boosting satellites into space was a tough challenge to crack; NASA was the organisation that provided the brainpower required. Easley was working in the Launch Vehicles Group at this time[3] and, with her skills in computer science and coding, contributed to the development of the Centaur upper-stage rocket. This rocket was used for more than 100 unmanned launches and led the way to understanding our solar system and planet, as well as enabling much of the satellite-reliant technology used by internet-dependent societies today.

Easley was keen to stress that her facility did not spring into existence because of the Space Race. The teams multi-tasked, working on a number of projects that included alternative power, energy conservation systems, and battery technology.[4] The hand she had in the life of storage batteries paved the way for the electric transportation that is likely to play a large role in tackling the climate crisis and determining the future of our world. Long before major car manufacturers started advertising the latest, most 'eco-friendly' cars of the future, Easley's team members were driving a little electric car around their facility.

As well as being good at her job during her thirty-four years at NASA, Easley was a huge advocate for equal rights in the workplace. Growing up, she remembered often being asked to stay behind after school for expressing her opinion in a way that was not appreciated by her teachers; as an adult, she was even less obliged to stay quiet when she witnessed such injustice around her. She became an Equal Employment Opportunity counsellor and interestingly observed that many people do not see a problem with lack of equality until they themselves feel that they are being discriminated against, for example because of their age.

Sadly, she had a few experiences that enabled her to empathise with the unfairness of those who came to her for counsel. One particularly bad example involved her being cropped out of a photograph that depicted her small team working on a piece of equipment. The photograph was enlarged and put on display at an open house that she attended with her colleagues, not knowing the humiliation that awaited her. Later on in her career, she decided to go back to university to study for a degree in maths. She did this whilst working and as a result had a punishing schedule, but the real issue revolved around payment. Even though the company had a history of supporting the undergraduate courses of other employees, Easley's supervisor was adamant the company would not contribute towards her course despite the fact that without the degree she would forever be classed as 'semi-professional' regardless of her competence and experience. Not one to back down, Easley funded her own education and earned the degree that would set her on a path to further success. It was only after she had handed in her grade that a colleague in the training department asked Easley's supervisor why she had not applied for financial aid, because it had been available all along. Although apologetic, it seems that Easley's supervisor never did anything to make amends or refund her for the cost of the course. To add further insult to injury, unlike other employees who held degrees, Easley was requested to attend additional training before she would be considered a professional.

It would be doing Easley a disservice to only focus on the time and energy she gave to her job and work environment. There is no doubt that she felt a huge sense of responsibility towards her community. This started as a young adult, during a time when the 'Jim Crow' laws required people in some states to pay a voting toll and take a literacy test. Because they suffered high levels of illiteracy due to increased barriers to education, these restrictions disproportionately affected black and Native Americans, minority ethnic groups, and all people of a lower economic class. Easley used her own intelligence to help others study for the literacy tests and thereby increase their chance of being able to vote. She continued tutoring even after moving to Cleveland and working full-time.

Easley had followed in her mother's footsteps; by encouraging participation in STEM and doing outreach, she took on the qualities of her own role model. Without a doubt, countless others were inspired, and went on to inspire because of her.

Further Reading

Velentina Tereshkova – a Russian engineer and cosmonaut who was the first woman in space.

Katherine Johnson – a mathematician who worked on calculations that led to the first US spaceflights.

Mary Jackson – an American aerospace engineer and mathematician who became NASA's first black female engineer.

Constance Tipper (1894–1995)
Marie Gayler (1891–1976)
Marion McQuillan (1921–1998)
Materials Science and Engineering
England

How long does it take for a crack in a steel beam to grow to the point where the bridge it is holding up fails catastrophically? How can the chemical composition and structure of natural materials like slug slime and spider silk provide cutting-edge solutions to repair the human body? How can ceramics play a role in dealing with nuclear waste?

A great deal of science and engineering hinges on a discipline that rarely receives the attention it deserves. Long before great structures can be designed and built, or before a surgeon implants a hip replacement in a patient suffering from life-limiting arthritis, a group of scientists and engineers have spent decades researching and developing the materials used. They understand the microscopic structure of the material that allows the behaviour to be predicted when it is subjected to different conditions, such as being placed under stress or exposed to different liquids. Using a range of STEM fields – which can include everything from maths, physics, and chemistry, to biology, physiology, and medicine – materials scientists have had a big hand in shaping the world we live in today.

Although this interdisciplinary field was born out of sciences that were traditionally male-dominated, many women have played pivotal roles in contributing to its development. As early as the twenty-seventh century BC, it is thought that the Chinese Empress Xi Ling-shi was responsible for discovering that fibres produced by silkworms could be woven into the strong and yet incredibly beautiful fabrics that have been valued throughout history.[1] Female innovators have also had success in more recent centuries, studying the materials around us and learning how they behave as well as finding useful applications for them. Constance Tipper was a Cambridge-educated metallurgist who specialised in crystallography, an area of science that studies how atoms in many materials are arranged in a regular three-dimensional pattern. At the Royal School of Mines, she was part of the team studying the process of how crystal structures change when a metal is deformed. Tipper researched the process of recrystallisation, during which atoms in a deformed metal rearrange themselves when subsequently reheated or 'annealed'[2] – this is a staple of metallurgy as it can completely alter the properties of the material, making it useful for some applications and dangerous for others. After moving to Cambridge, she started working with Geoffrey Taylor, who built on the research they did together to develop the concept of dislocations,[3] which shows that irregularity within the crystal structure of a metal has an impact on the material properties. They applied tensile stress to aluminium and investigated the distortion of atoms within the crystal structure. This work earned them an invitation from the Royal Society to deliver the prestigious Bakerian lecture, though it was hastily retracted from Tipper once they realised she was a woman.[4] Despite this (albeit retracted) honour, Tipper only received an official title in the engineering department at the onset of the Second World War. Her promotion was a wise decision on the university's part because the War Office would shortly be in need of her expertise to solve a problem that was hindering the success of allied forces. Thousands of vessels called Liberty Ships were mass-produced by the US to transport cargo and personnel during the war, but there had been multiple instances of cracking in the hull and deck. In some extreme cases, the ships split in half. The new head of the engineering department at Cambridge had been considering the welding between the plates as a point of failure, but Tipper put forward the idea that overloading and the cold conditions of the north Atlantic were causing the metal to become brittle, allowing cracks to propagate and eventually cause fractures.[5] In her efforts to provide quantitative results, she devised an experimental test that determines the brittleness of structural mild steel at service temperature.

Born just a few years before Constance Tipper, Marie Gayler was another metallurgist who was a pioneer in her field. Both women presented papers at the 1920 meeting organised by the Institute of Metals,[6] but their areas of research were quite different. Gayler, who had a background in chemistry, worked at the National Physical Laboratory for most of her career and is remembered for her work on alloys. There are several processing mechanisms that can change the properties of metal, for example to increase strength. Gayler and her colleagues used a heat treatment on a specific aluminium alloy to increase the yield strength. This work was later built on to develop the Y alloy which could be used inside internal combustion engines. It was also used on the outside of Concorde, which needed to withstand higher temperatures than other aircraft due to its speed.[7]

Although difficult to imagine in the twenty-first-century world of work, Gayler's marriage to fellow metallurgist John Haughton could well have put an end to her career at the National Physical Laboratory. At the time, the marriage bar required women to leave the civil service when they got married. Fortunately, she was able to receive a waiver that gave her permission to keep her position which was incredibly rare. From the mid-1930s, Gayler's professional interests turned to dental amalgams. She tested a range of alloys using different analysis techniques to see how the amalgam would set[8] – research that paved the way for the future development of dental materials that are vital for improving quality of life.

Another important material, titanium, is a well-researched metal nowadays, but this was not always the case. It was first discovered in 1791, but it is thanks to pioneers like Marion McQuillan that its remarkable properties were researched and shared with the global science community.

After graduating from the University of Cambridge, McQuillan's first job was at the Royal Aircraft Establishment, Farnborough in 1942.

As with many innovations in science and engineering, war and international rivalries played a strong part. Titanium was first experimented on by both the Americans and the Russians for use in military jets.[9] The benefit of titanium over steel is its higher strength to weight ratio.[10] This means that the same part can be made the same size with the same strength but lower weight. This is crucial when it comes to aircraft as every kilogram saved leads to an improvement in manoeuvrability, range, or maximum payload; all crucial factors when fighting a war.

The use of titanium eventually trickled its way down to the commercial aircraft industry. Where the military favoured benefits that could be useful during war, the benefits to commercial airlines were primarily based on financial implications. Titanium is more expensive than steel for the same component as both the manufacturing of the raw material and the process of forming it into an object are more difficult. However, the weight saving allowed for less fuel to be used in flight and an increase in the cargo and number of passengers that an aircraft could carry. Over the course of an aircraft's life, both of these factors would increase revenue enough to offset the higher price of the titanium components. This shift had a part to play in connecting global economies and cultures as lower fares meant that air travel became more affordable.

The lives of these three women are an inspiring example of how different areas of science can be blended together to solve problems and shape entire industries. The challenge for the future is to continue the development of materials that contribute to a more sustainable way of life for all, not just for those who can afford cutting-edge technology. Advancements in material science drive technological innovation; the results of the manipulation of material structures and therefore properties are clear to see in daily life. For example, graphene has advanced screens of the smartphones many of us have come to rely on, and improvements in battery technology continue to bring desperately needed improvements in energy efficiency. By studying materials in nature and mimicking them, we can design materials that stand the test of time and often work symbiotically with the earth rather than against it. There are still many challenges to overcome; these women and their peers laid the foundations, but it is up to future generations of scientists and engineers to build upon and modify them for the ever-evolving requirements of our modern world. The achievements of these women and the work that still needs to be done to solve current and future challenges are intrinsically linked. Women today can see further in the world of science and engineering not because they are taller but because they are already standing on the shoulders of those giants who went before them.

Notes

Introduction
1. C. C. Perez, *Invisible Women: Exposing Data Bias in a World Designed for Men*, Random House, 2019.
2. National Highway Traffic Safety Administration, 'Injury Vulnerability and Effectiveness of Occupant Protection Technologies for Older Occupants and Women', US Department of Transportation, 2013.

Chapter 1: World Defining Moments in History
Joan Clarke
1. S. Budiansky, *Battle of Wits: The Complete Story of Codebreaking in World War II*, New York: The Free Press, 2000.
2. Lord Stewartby, 'Obituary Mrs J.E.L. Murray', *British Numismatic Society*, vol. 67, no. 13, pp. 162–6, 1997.
3. 'The Rising Tide: Women at Cambridge', University of Cambridge, [Online]. Available: https://www.cam.ac.uk/TheRisingTide. [Accessed 12 February 2021].
4. A. Stripp, 'Chapter 8: Hut 8 and naval Enigma, Part I', *Codebreakers: The Inside Story of Bletchley Park*, Oxford University Press, 2001, p. 113.
5. F. Carter, 'The Turing Bombe', *Rutherford Journal*, [Online]. Available: http://www.rutherfordjournal.org/article030108.html. [Accessed 12 February 2021].
6. A. Stripp, 'Chapter 8: Hut 8 and naval Enigma, Part I', *Codebreakers: The Inside Story of Bletchley Park*, Oxford University Press, 2001, p. 114.
7. C. Grey, Decoding Organization: Bletchley Park, Codebreaking and Organization Studies, Cambridge: Cambridge University Press, 2012.
8. Z. Tsjeng, 'Mathematics, The Bletchleyettes', *Forgotten Women: The Scientists*, Octopus Publishing Group, 2018, p. 163.
9. C. Grey, *Decoding Organization: Bletchley Park, Codebreaking and Organization Studies*, Cambridge University Press, 2012.

Rosalind Franklin
1. 'Rosalind Franklin: Biographical Overview', US National Library of Medicine, [Online]. Available: https://profiles.nlm.nih.gov/spotlight/kr/feature/biographical. [Accessed 03 March 2021].
2. 'Rosalind Franklin: Biographical Overview', US National Library of Medicine, [Online]. Available: https://profiles.nlm.nih.gov/spotlight/kr/feature/biographical. [Accessed 11 March 2021].
3. 'Maurice Wilkins – Nobel Lecture', Nobel Media, [Online]. Available: https://www.nobelprize.org/prizes/medicine/1962/wilkins/lecture/. [Accessed 15 March 2021].

4. J. Watson, *The Double Helix: A Personal Account of the Discovery of the Structure of DNA*, Simon & Schuster, 1968.
5. M. Cobb, 'Sexism in science: did Watson and Crick really steal Rosalind Franklin's data?' *The Guardian*, 23 Jun 2015. [Online]. Available: https://www.theguardian.com/science/2015/jun/23/sexism-in-science-did-watson-and-crick-really-steal-rosalind-franklins-data. [Accessed 23 March 2021].
6. 'Maurice Wilkins – Nobel Lecture', *Nobel Media*, [Online]. Available: https://www.nobelprize.org/prizes/medicine/1962/wilkins/lecture/. [Accessed 25 March 2021].
7. M. Cobb, *Life's Greatest Secret: The Race to Crack the Genetic Code*, Profile Books, 2015.
8. R. E. Franklin and R. G. Gosling, 'The structure of sodium thymonucleate fibres. I. The influence of water content', *Acta Crystallographica*, vol. 6, pp. 673–7, 1953.
9. M. Cobb, *Life's Greatest Secret: The Race to Crack the Genetic Code*, Profile Books, 2015.
10. M. Cobb, 'Sexism in science: did Watson and Crick really steal Rosalind Franklin's data?' *The Guardian*, 23 June 2015. [Online]. Available: https://www.theguardian.com/science/2015/jun/23/sexism-in-science-did-watson-and-crick-really-steal-rosalind-franklins-data. [Accessed 11 March 21].
11. J. Watson and F. Crick, 'Molecular Structure of Nucleic Acids: A Structure for Deoxyribose Nucleic Acid', *Nature*, vol. 171, pp. 737–8, 1953.
12. 'Rosalind Franklin, Virus Researcher', *New York Times*, p. 85, 20 April 1958.

Ada Lovelace
1. B. Wooley, *The Bride of Science: Romance, Reason and Byron's Daughter*, Macmillan, 1999.
2. 'Difference Engine No.2, designed by Charles Babbage, built by Science Museum', Science Museum Group, [Online]. Available: https://collection.sciencemuseumgroup.org.uk/objects/co526657/difference-engine-no-2-designed-by-charles-babbage-built-by-science-museum-difference-engine. [Accessed 02 January 2021].
3. B. A. Toole, *Ada, the Enchantress of Numbers*, Strawberry Press, 1998.
4. D. Swade, 'Charles Babbage and Difference Engine No. 2 | Doron Swade | Talks at Google', Talks at Google, 12 May 2008. [Online]. Available: https://youtu.be/7K5p_tBcrd0. [Accessed 02 January 2020].
5. B. Klare, M. Burge, J. Klontz, R. Vorder Bruegge and A. Jain, 'Face Recognition Performance: Role of Demographic Information', IEEE Transactions on Information and Security, vol. 7, no. 6, p. 1789, 2012.

Lise Meitner
1. P. Rife, Lise Meitner and the Dawn of the Nuclear Age, Plunkett Lake Press, 2017.
2. ibid.
3. ibid.
4. J. Chadwick, 'Existence of a Neutron', *Proceedings of the Royal Society A.*, vol. 136, no. 830, pp. 692–708, 1932.
5. 'Otto Hahn – Nobel Lecture', *Nobel Media*, [Online]. Available: https://www.nobelprize.org/prizes/chemistry/1944/hahn/lecture/. [Accessed 14 March 2021].

Rufaida Al-Aslamia
1. K. Miller-Rosser, Y. Chapman and K. Francis, 'Historical, Cultural, and Contemporary Influences on the Status of Women in Nursing in Saudi Arabia', OJIN: The Online Journal of Issues in Nursing, vol. 11, no. 3, 2006.
2. H. Jawad, *The Rights of Women in Islam: An Authentic Approach*, Palgrave, 1998.
3. R. J. Rukanuddin, 'Rufaida Al-Asalmiya, the First Muslim Nurse', *Journal of Nursing Scholarship*, vol. 28, no. 3, pp. 267–78, 1996.
4. S. Lovering, 'The Crescent of Care: A nursing model to guide the care of Arab Muslim patients', *Diversity and Equality in Health and Care*, vol. 9, no. 3, pp. 171–8, 2012.
5. K. Miller-Rosser, Y. Chapman and K. Francis, 'Historical, Cultural, and Contemporary Influences on the Status of Women in Nursing in Saudi Arabia', OJIN: The Online Journal of Issues in Nursing, vol. 11, no. 3, 2006.

Chapter 2: Our Day-to-Day Lives
1. M. J. Gage, 'Woman as an Inventor', *North American Review*, vol. 136, no. 318, p. 488, 1883.
2. UK Intellectual Property Office Informatics Team, Gender Profiles in Worldwide Patenting: An analysis of female inventorship, Intellectual Property Office, 2016.
3. ibid. (2019 Edition), Intellectual Property Office, 2019.
4. ibid.
5. ibid. Intellectual Property Office, 2016.

Alice Parker
1. United States Census Bureau, United States Census, 1920, United States Census Bureau, 1920.
2. ibid.

Josephine Cochrane
1. E. Sobey, *The Way Kitchens Work: The Science Behind the Microwave, Teflon Pan, Garbage Disposal, and More*, Chicago Review Press, 2010.
2. J. Ram and E. Atkisson, 'I'll Do It Myself', United States Patent and Trademark Office, [Online]. Available: https://www.uspto.gov/learning-and-resources/journeys-innovation/historical-stories/ill-do-it-myself. [Accessed 13 January 2021].
3. J. Cochran, 'Washing or rinsing machines for crockery or tableware with stationary crockery baskets and spraying devices within the cleaning chamber with rigidly mounted spraying devices', United States Patent 355139, 28 December 1886.
4. J. Ram and E. Atkisson, 'I'll Do it Myself', United States Patent and Trademark Office, [Online]. Available: https://www.uspto.gov/learning-and-resources/journeys-innovation/historical-stories/ill-do-it-myself. [Accessed 13 January 2021].
5. ibid.

Mary Anderson
1. 'Mary Anderson: Windshield Wiper',' Massachusetts Institute of Technology School of Engineering, [Online]. Available: https://lemelson.mit.edu/resources/mary-anderson. [Accessed 22 March 2021].

Mary Beatrice Davidson Kenner
1. D. L. Davis, 'Women Inventors: Lest we Forget', NCPedia, 2006. [Online]. Available: https://ncpedia.org/industry/women-inventors. [Accessed 28 March 2021].
2. A. Hambrick, Biographies of black female scientists and inventors: an interdisciplinary middle school curriculum guide: 'What shall I tell my children who are black?' ScholarWorks@UMass Amherst, 1993.
3. ibid.
4. P. C. Sluby, 'African American Brilliance', *Tar Heel Junior Historian*, vol. 46, no. 1, 2006.
5. L. S. Jeffrey, *Amazing American Inventors of the 20th Century*, Enslow Publishers Inc, 2013.
6. A. Hambrick, Biographies of black female scientists and inventors: an interdisciplinary middle school curriculum guide: 'What shall I tell my children who are black?' ScholarWorks@UMass Amherst, 1993.
7. ibid.
8. ibid.

Stephanie Kwolek
1. 'Stephanie Kwolek (1923–2004)', American Chemical Society, [Online]. Available: www.acs.org/content/acs/en/education/whatischemistry/women-scientists/stephanie-kwolek.html. [Accessed 09 January 2021].
2. 'IACP/DUPONT™ KEVLAR® SURVIVORS' CLUB', Du Pont, [Online]. Available: https://www.dupont.com/military-law-enforcement-and-emergency-response/the-kevlar-survivors-club.html. [Accessed 09 January 2021].

Ruby Hirose
1. A. Lommen and K. Lommen, 'Hirose Family', Auburn Pioneer Cemetery, [Online]. Available: http://auburnpioneercemetery.net/biographies/hirose.php. [Accessed 19 February 2021].
2. Interview with Ruby Hirose, Stanford Survey on Race Relations, 1924–1927.
3. ibid.
4. US Government, Naturalization Act of 1870, 16 Stat, Chapter 254, US Government, 1870.
5. Interview with Ruby Hirose, Stanford Survey on Race Relations, 1924–1927.
6. R. Hirose, 'The Second Phase of Thrombin Action: Fibrin Resolution', *American Journal of Physiology*, vol. 107, no. 3, pp. 693–7, 1934.
7. ibid.
8. University of Cincinnati Commencement Program, University of Cincinnati, 1931.
9. 'Poliomyelitis', WHO, 22 July 2019. [Online]. Available: https://www.who.int/news-room/fact-sheets/detail/poliomyelitis. [Accessed 20 February 2021].
10. A. T. Glenny and M. Barr, 'The precipitation of diphtheria toxoid by potash alum', *The Journal of Pathology and Bacteriology*, vol. 34, no. 2, pp. 131–8, 1931.
11. 'Ruby Hirose: Lauded Japanese American scientist', Smithsonian, [Online]. Available: https://womenshistory.si.edu/herstory/object/ruby-hirose. [Accessed 20 February 2021].
12. 'Ruby Hirose: Lauded Japanese American scientist', Smithsonian, [Online]. Available: https://womenshistory.si.edu/herstory/object/ruby-hirose. [Accessed 20 February 2021].

13. M. Mbow, E. De Gregorio and J. Ulmer, 'Alum's adjuvant action: grease is the word', *Nature Medicine*, vol. 17, pp. 415–16, 2011.
14. 'Document for February 19th: Executive Order 9066: Resulting in the Relocation of Japanese', National Archives, [Online]. Available: https://www.archives.gov/historical-docs/todays-doc/?dod-date=219. [Accessed 20 February 2021].
15. Z. Tsjeng, *Forgotten Women: The Scientists*, Octopus Books, 2018.
16. 'Increasing Need Seen for Women Chemists', *The Cincinnati Enquirer*, p. 11, 18 March 1940.
17. A. Lommen and K. Lommen, 'Hirose Family', Auburn Pioneer Cemetery, [Online]. Available: http://auburnpioneercemetery.net/biographies/hirose.php. [Accessed 19 February 2021].

Olive Dennis
1. J. Padron, 'Olive Dennis – Innovating the Passenger Experience', National Railroad Museum, [Online]. Available: https://nationalrrmuseum.org/blog/olive-dennis-innovating-the-passenger-experience/. [Accessed 28 March 2021].
2. 'Friends of the Goucher Library Presents: Executive Influence: Olive Wetzel Dennis', Goucher College, 17 April 2012. [Online]. Available: https://web.archive.org/web/20120526090521/http://www.goucher.edu/x46553.xml. [Accessed 28 March 2021].
3. E. Engst and H. R. Segelken, 'Cornell Rewind: Phenomenal first women of engineering', *Cornell Chronicle*, 18 February 2015.
4. 'The Moving Assembly Line and Five Dollar Work Day', [Online]. Available: https://corporate.ford.com/articles/history/moving-assembly-line.html. [Accessed 28 March 2021].
5. S. E. Hatch, *Changing Our World: True Stories of Women Engineers*, American Society of Civil Engineers, 2006.
6. J. Padron, 'Olive Dennis – Innovating the Passenger Experience', National Railroad Museum, [Online]. Available: https://nationalrrmuseum.org/blog/olive-dennis-innovating-the-passenger-experience/. [Accessed 28 March 2021].
7. C. Giaimo, 'The 'Lady Engineer' Who Took the Pain Out of the Train', Atlas Obscura, 9 April 2018. [Online]. Available: https://www.atlasobscura.com/articles/olive-dennis-train-comfort-engineer. [Accessed 28 March 2021].
8. E. Engst and H. R. Segelken, 'Cornell Rewind: Phenomenal first women of engineering', *Cornell Chronicle*, 18 February 2015.

Sutayta Al-Mahāmali
1. V. J. Katz, 'Stages in the history of Algebra with Implications for Teaching', *Educational Studies in Mathematics*, vol. 66, pp. 185–201, 2007.
2. J. L. Esposito, *The Oxford History of Islam*, Oxford University Press, 2000.
3. P. Maher, 'From Al-Jabr to Algebra', *Mathematics in School*, vol. 27, no. 4, p. 14, 1998.
4. 'Telling SAGA: Improving Measurement and Policies for Gender Equality in Science, Technology and Innovation', SAGA Working Paper 5, UNESCO, p. 28, 2018.
5. T. Alkiek, 'Sayedaty Ep. 6: Sutayta Al-Mahāmali', Yaqeen Insitute, 13 August 2017. [Online]. Available: https://yaqeeninstitute.org/tesneem-alkiek/sayedaty-ep-6-sutayta-al-mahamali. [Accessed 09 March 2021].

6. D. Debakcsy, 'The Algebraist of Baghdad: Sutayta Al-Mahāmali's Medieval Mathematics', Women You Should Know, 1 November 2017. [Online]. Available: https://womenyoushouldknow.net/sutayta-al-mahamalis-mathematics/. [Accessed 09 March 2021].

Chapter 3: Improving Lives, Equality and Justice
Margaret Sanger
1. 'Margaret Sanger, 'What Every Girl Should Know', *New York Call*, p. 15, 2 March 1913.
2. 'The History and Impact of Planned Parenthood', Planned Parenthood, [Online]. Available: https://www.plannedparenthood.org/about-us/who-we-are/our-history. [Accessed 19 March 2021].
3. C. R. McCann, *Birth control politics in the United States, 1916-1945*, Cornell University Press, 1994.
4. ibid.
5. 'Nation's Premier Civil Rights Organization', National Association for the Advancement of Colored People, [Online]. Available: https://naacp.org/nations-premier-civil-rights-organization/. [Accessed 19 March 2021].
6. C. McCann, *Birth control politics in the United States, 1916-1945*, Cornell University Press, 1994.
7. 'The Reverend Martin Luther King Jr. Upon Accepting the Planned Parenthood Sanger Award', Planned Parenthood, 2004. [Online]. Available: https://www.plannedparenthood.org/planned-parenthood-gulf-coast/mlk-acceptance-speech. [Accessed 19 March 2021].
8. J. Carey, 'The Racial Imperatives of Sex: Birth Control and Eugenics in Britain, the United States and Australia in the interwar years', *Women's History Review*, vol. 21, no. 5, pp. 733–52, 2012.
9. D. Barrett and C. Kurzman, 'Globalizing Social Movement Theory: The Case of Eugenics', *Theory and Society*, vol. 33, no. 5, pp. 487–527, 2004.
10. M. Hawkins, *Social Darwinism in European and American Thought*, Cambridge University Press, 1997.
11. J. Carey, 'The Racial Imperatives of Sex: Birth Control and Eugenics', *Women's History Review*, vol. 21, no. 5, pp. 733–52, 2012.
12. A. Franks, *Margaret Sanger's Eugenic Legacy*, McFarland & Company Inc., Publishers, 2005.
13. M. Sanger, 'Letter from Margaret Sanger to Dr. C. J. Gamble, December 10, 1939', Smith Libraries Exhibits, [Online]. Available: https://libex.smith.edu/omeka/items/show/495. [Accessed 19 March 2021].
14. 'Birth Control or Race Control? Sanger and the Negro Project', The Margaret Sangers Papers Project, Fall 2001. [Online]. Available: http://www.nyu.edu/projects/sanger/articles/bc_or_race_control.php. [Accessed 19 March 2021].
15. 'Social Unemployed', New York University, Margaret Sanger Project, 2003. [Online]. [Accessed 19 March 2021].
16. N. Stewart, 'Planned Parenthood in N.Y. Disavows Margaret Sanger Over Eugenics', *The New York Times*, 21 July 2020. [Online]. Available: https://www.nytimes.com/2020/07/21/nyregion/planned-parenthood-margaret-sanger-eugenics.html. [Accessed 19 March 2021].

17. J. Lepore, 'The Surprising Origin Story of Wonder Woman', *Smithsonian Magazine*, October 2014. [Online]. Available: https://www.smithsonianmag.com/arts-culture/origin-story-wonder-woman-180952710/. [Accessed 19 March 2021].

Anne McLaren
1. S. Franklin, 'Obituary: Dame Dr Anne McLaren', *Regenerative Medicine*, vol. 2, no. 5, pp. 853–9, 2007.
2. M. Hargittai, *Women Scientists: Reflections, Challenges, and Breaking Boundaries*, Oxford University Press, 2015.
3. S. Franklin, 'Obituary: Dame Dr Anne McLaren', *Regenerative Medicine*, vol. 2, no. 5, pp. 853–9, 2007.
4. 'Preface', Special issue dedicated to honouring the career of Anne McLaren, International Journal of Developmental Biology, *Mammalian reproduction and development*, vol. 45, no. 3, pp. 454–622, 2001.
5. M. Hargittai, *Women Scientists: Reflections, Challenges, and Breaking Boundaries*, Oxford University Press, 2015.
6. S. Franklin, 'Obituary: Dame Dr Anne McLaren', *Regenerative Medicine*, vol. 2, no. 5, pp. 853–9, 2007.
7. M. Hargittai, *Women Scientists: Reflections, Challenges, and Breaking Boundaries*, Oxford University Press, 2015.
8. The Warnock Committee, *British Medical Journal*, vol. 289, pp. 238–9, 1984.
9. S. Franklin, 'Obituary: Dame Dr Anne McLaren', *Regenerative Medicine*, vol. 2, no. 5, pp. 853–9, 2007.

Laura Bassi
1. P. Findlen, 'Science as a Career in Enlightenment Italy: The Strategies of Laura Bassi', *ISIS*, vol. 84, no. 3, pp. 441–69, 1993.
2. A. Elena, 'In lode della filosofessa di Bologna: an introduction to Laura Bassi', *ISIS*, vol. 82, pp. 510–18, 1991.
3. ibid.
4. ibid.
5. P. Findlen, 'Laura Bassi and the city of learning', Physics World, 29 August 2013. [Online]. Available: https://physicsworld.com/a/laura-bassi-and-the-city-of-learning/. [Accessed 18 September 2020].

Alice Gullattee
1. 'Interview with Dr Alyce Gullattee', *Journal of the Howard University Institute on Drug Abuse and Addiction*, vol. 2, no. 1, pp. 4–7, 1972.
2. Berkeley, University of California, Register – University of California, Volume 2, Berkeley, CA: University of California Press, 1957.
3. 'Interview with Dr Alyce Gullattee', *Journal of the Howard University Institute on Drug Abuse and Addiction*, vol. 2, no. 1, pp. 4–7, 1972.
4. 'SNMA 50th Anniversary Video', The Student National Medical Association, 26 March 2015. [Online]. Available: https://snma.org/page/history. [Accessed 21 March 2021].
5. 'Interview with Dr Alyce Gullattee', *Journal of the Howard University Institute on Drug Abuse and Addiction*, vol. 2, no. 1, pp. 4–7, 1972.

6. Unknown, 'In Huddle on Human Kindness', *Washington Afro American*, p. 6, 20 May 1975.
7. C. T. Rowan, 'Heroin and Tragedy at a Very Early Age', *The Cincinnati Enquirer*, p. 13, 26 March 1981.
8. 'Alice Gullattee', African American Registry, [Online]. Available: https://aaregistry.org/story/alyce-gullattee-born/. [Accessed 22 March 2021].

Chapter 4: Global Health
James Barry
1. M. du Preez and J. Dronfield, Dr James Barry: A Woman Ahead of Her Time, Oneworld Publications, 2016.
2. M. Gelfand, 'The Somerset tradition: Dr James Barry', *South Africa Medical Journal*, vol. 39, no. 32, pp. 511–14, 1965.
3. H. M. du Preez, 'Dr James Barry: The Early Years Revealed', *South African Medical Journal*, vol. 98, no. 1, 2008.

Patricia Bath
1. A. Green, 'Obituary', *The Lancet*, vol. 394, no. 10197, p. 464, 2019.
2. 'Teenage Scientist Is Named One of the Ten Young Women of The Year', *Atlanta Daily World*, 31 December 1960.
3. L. Lambert, 'Patricia Bath: Inventor of laser cataract surgery', *Inventors and Inventions*, Marshall Cavendish, p. 70, 2007.
4. M. Davidson, 'Innovative Lives: The Right To Sight: Patricia Bath', Smithsonian National Museum of American History, Lemelson Center for the Study of Invention and Innovation, 3 March 2005. [Online]. Available: https://invention.si.edu/innovative-lives-right-sight-patricia-bath . [Accessed 12 December 2020].
5. United States Census Bureau, Persons by Poverty Status, by State: 1959, 1969, 1979, 1989, 1999, United States Census Bureau.
6. A. Green, 'Obituary', *The Lancet*, vol. 394, no. 10197, p. 464, 2019.
7. Conversation Between Patricia Bath, MD and Eve Higginbotham, MD, The Foundation of the American Academy of Ophthalmology Museum of Vision & Ophthalmic Heritage, Orlando, 2011.
8. P. E. Bath, 'Rationale for a Program in Community Ophthalmology', Journal of the National Medical Association, vol. 71, no. 2, pp. 145-148, 1979.
9. M. Davidson, 'Innovative Lives: The Right To Sight: Patricia Bath', Smithsonian National Museum of American History, Lemelson Center for the Study of Invention and Innovation, 3 March 2005. [Online]. Available: https://invention.si.edu/innovative-lives-right-sight-patricia-bath. [Accessed 12 December 2020].
10. 'Patricia Bath On Being The First Person To Invent & Demonstrate Laserphaco Cataract Surgery | TIME,' *Time Magazine*, 30 October 2017. [Online]. Available: https://www.youtube.com/watch?v=gcE_QMTBNW4. [Accessed 13 December 2020].
11. National Library of Medicine, 'Dr Patricia E. Bath', National Library of Medicine, [Online]. Available: https://cfmedicine.nlm.nih.gov/physicians/biography_26.html. [Accessed 18 December 2020].
12. 'American Institute for the Prevention of Blindness', American Institute for the Prevention of Blindness, [Online]. Available: http://www.blindnessprevention.org/about.php. [Accessed 18 December 2020].

13. 'Patricia Bath–Laserphaco Probe', Lemelson MIT Programme, Massachusetts Institute of Technology, [Online]. Available: https://lemelson.mit.edu/resources/patricia-bath. [Accessed 18 December 2020].
14. N. Genzlinger, 'Dr. Patricia Bath, 76, Who Took On Blindness and Earned a Patent, Dies', *New York Times*, 4 June 2019.
15. M. Davidson, 'Innovative Lives: The Right To Sight: Patricia Bath', Smithsonian National Museum of American History, Lemelson Center for the Study of Invention and Innovation, 3 March 2005. [Online]. Available: https://invention.si.edu/innovative-lives-right-sight-patricia-bath. [Accessed 12 December 2020].
16. Career Communications Group, '1997 Women of Color', in *US Black Engineer & IT*, 1997, p. 42.
17. 'Patricia Bath On Being The First Person To Invent & Demonstrate Laserphaco Cataract Surgery | TIME', *Time Magazine*, 30 October 2017. [Online]. Available: https://www.youtube.com/watch?v=gcE_QMTBNW4. [Accessed 13 December 2020].
18. 'Conversation Between Patricia Bath, MD and Eve Higginbotham, MD', The Foundation of the American Academy of Ophthalmology Museum of Vision & Ophthalmic Heritage, Orlando, FL, 2011.

Kin Yamei

1. W. Shurtleff and A. Aoyagi, 'Biography of Yamei Kin MD (1864-1934), (Also Known as Jin Yunmei), the first Chinese Woman to Take a Medical Degree in the United States (1864-2016)', Soyinfo Centre, 2016.
2. ibid.
3. W. A. N. Clara, *Clara's diary: an American girl in Meiji Japan*, Tokyo, New York, San Francisco, Kodansha International Ltd, p. 353, 1979.
4. 'M. King', *Sumner Gazette*, p. 1, 11 June 1885.
5. W. Shurtleff and A. Aoyagi, Biography of Yamei Kin MD (1864-1934), (Also Known as Jin Yunmei), the first Chinese Woman to Take a Medical Degree in the United States (1864-2016), Soyinfo Centre, 2016.
6. N. Y. Shizuka and M. Kin, 'A young Chinese medical missionary in Kobe: A sketch of the former half of her life', *Jinbun Ronkyu (Humanities Review)*, vol. 3, no. 48, pp. 174-188, 1998.
7. M. Roth, 'The Chinese-Born Doctor Who Brought Tofu to America', *Smithsonian Magazine*, 13 August 2018. [Online]. Available: https://www.smithsonianmag.com/history/chinese-born-doctor-who-brought-tofu-america-180969977 [Accessed 04 February 2021].
8. 'Anxious to be a 'new' woman: Chinese wife adopts American customs and deserts her spouse to become a doctor. Now lives in Boston and supports herself. Man with an extensive name, who was married to her in Orient, gets a divorce–Other unhappy marriages', *San Francisco Chronicle*, p. 10, 13 August 1904.
9. 'Letters tell of love and plans BIOGRAPHY OF YAMEI KIN MD', *San Francisco Chronicle*, p. 10, 13 September 1904.
10. F. G. Carpenter, 'Medical College for Women', *Chicago Daily Tribune*, August 1909.
11. 'Forty young Chinese women are qualifying for the medical profession in American universities', *Chicago Commerce*, 10 May 1912.
12. 'Woman off to China as government agent to study soy bean', *New York Times Magazine*, p. 9, 10 June 1917.

13. J. Prusek, *My Sister China*, Prague: The Karolinium Press, 2002.
14. Wong and Wu, *History of Chinese Medicine*, pp.557–8, 1936.
15. J. Prusek, *My Sister China*, Prague: The Karolinium Press, 2002.

Kamala Sohonie
1. S. Sarkar, *Modern India 1885-1947*, Macmillan Publishers India, 1983.
2. A. Mitra, 'The Life and Times of Kamala Bhagvat Sohonie; The Unsung Hero of Science in India', *Resonance*, vol. 21, no. 4, p. 304, 2016.
3. A. Gupta, 'Kamala Sohonie', [Online]. Available: https://www.arvindguptatoys.com/arvindgupta/bs28ksohonie.pdf. [Accessed 20 April 2020].
4. C. V. Raman, 'On the Molecular Scattering of Light in Water and the Colour of the Sea', *Proceedings of The Royal Society*, vol. 101, no. 708, pp. 64–80, 1922.
5. A. Gupta, 'Bright Sparks: Inspiring Indian scientists from the past', [Online]. Available: https://www.arvindguptatoys.com/arvindgupta/bsintro.pdf. [Accessed 21 April 2020].
6. V. Dhuru, 'The scientist lady', [Online]. Available: https://www.ias.ac.in/public/Resources/Initiatives/Women_in_Science/Contributors/kamalasohonie.pdf. [Accessed 21 April 2020].
7. D. Chattopadhyay, 'Kamala Sohonie: First Indian woman PhD in science', *Science and Culture*, vol. 81, pp. 128–30, 2015.
8. ibid.
9. K. Bhagvat and M. Sreenivasaya, 'The non-protein nitrogen of pulses', *Biochemical J.*, vol. 29, pp. 909-913, 1935.
10. A. Mitra, 'The Life and Times of Kamala Bhagvat Sohonie; The Unsung Hero of Science in India', *Resonance*, vol. 21, no. 4, p. 304, 2016.
11. K. Nickelsen, *Explaining Photosynthesis: Models of Biochemical Mechanisms, 1840–1960*, Springer, 2015.
12. D. Chattopadhyay, 'Kamala Sohonie: First Indian woman PhD in science', *Science and Culture*, vol. 81, pp. 128–30, 2015.
13. I. Talbot and S. Gurharpal, *The Partition of India*, Cambridge University Press, 2009.
14. D. Chattopadhyay, 'Kamala Sohonie: First Indian woman PhD in science', *Science and Culture*, vol. 81, pp. 128–30, 2015.
15. A. Mitra, 'The Life and Times of Kamala Bhagvat Sohonie; The Unsung Hero of Science in India', *Resonance*, vol. 21, no. 4, p. 304, 2016.
16. D. Richter, 'Opportunities for Women in India, Women Scientists', *Woman Scientists: The Road to Liberation*, Palgrave, 1982, p. 14.

Jane Cooke Wright
1. W. M. Cobb, 'Louis Tompkins Wright, 1891–1952', *Journal of the National Medical Association*, vol. 45, no. 2, pp. 130-148, 1953.
2. P. P. Reynolds, 'Dr. Louis T. Wright and the NAACP: Pioneers in Hospital Racial Integration', *American Journal of Public Health*, vol. 90, no. 6, pp. 883-892, 2000.
3. 'Meharry Medical College: About', Meharry Medical College, [Online]. Available: https://home.mmc.edu/about/. [Accessed 19 February 2021].
4. S. M. Swain, 'A Passion for Solving the Puzzle of Cancer: Jane Cooke Wright, MD, 1919–2013', *The Oncologist*, vol. 18, no. 6, pp. 646–8, 2013.
5. ibid.

6. J. C. Wright, J. I. Plummer, R. S. Coidan and L. T. Wright, 'In vivo and in vitro effects of chemotherapeutic agents on human neoplastic diseases', *Harlem Hospital Bulletin*, vol. 6, no. 2, pp. 58–63, 1953.
7. 'Paying Tribute to ASCO Founder Jane C. Wright, MD', Conquer Cancer Foundation, 21 June 2011. [Online]. Available: https://www.youtube.com/watch?v=6hHiWeki9GE. [Accessed 23 February 2021].
8. J. C. Wright, A. Prigot, B. Wright, S. Weintraub and L. T. Wright, 'An evaluation of folic acid antagonists in adults with neoplastic diseases: a study of 93 patients with incurable neoplasms', *Journal of the National Medical Association*, vol. 43, no. 4, pp. 211–40, 1951.
9. 'World Health Organization model list of essential medicines', WHO, 2019. [Online]. Available: https://apps.who.int/iris/handle/10665/325771. [Accessed 21 February 2021].
10. 'Jane Cooke Wright MD', American Association for Cancer Research, [Online]. Available: https://www.aacr.org/professionals/membership/aacr-academy/fellows/jane-cooke-wright-md/. [Accessed 23 February 2021].
11. J. C. Wright, J. P. Cobb, S. L. Gumport, D. Safadi, D. G. Walker and F. M. Golomb, 'Further investigation of the relation between the clinical and tissue culture response to chemotherapeutic agents on human cancer', *Cancer*, vol. 15, pp. 284–93, 1962.
12. 'Jane Cooke Wright MD', American Association for Cancer Research, [Online]. Available: https://www.aacr.org/professionals/membership/aacr-academy/fellows/jane-cooke-wright-md/. [Accessed 23 February 2021].
13. H. L. Crosby, 'Jane Cooke Wright (1919–2013): Pioneering Oncologist, Woman and Humanitarian.', *Journal of Medical Biography*, vol. 24, no. 1, pp. 38-41, 2016.

Alice Ball
1. D. M. Guttman and E. Golden, *African Americans in Hawai'i*, Arcadia Publishing, p. 15, 2011.
2. H. T. Hollman, 'The Fatty Acids of Chaulmoogra Oil in the Treatment of Leprosy and Other Diseases', *Archives of Dermatology*, vol. 5, no. 1, pp. 94–101, 1922.
3. P. Wermager and C. Heltzel, 'Alice A. Augusta Ball: Young Chemist Gave Hope to Millions', *ChemMatters* (American Chemical Society), p. 17, 2007.
4. A. L. Dean and R. Wrenshall, 'Fractionation of Chaulmoogra Oil', *Journal of the American Chemical Society*, vol. 42, no. 12, pp. 2626-2645, 1920.
5. S. Kreifels, 'Alice Ball made a stunning find in her early 20s', *Honolulu Star-Bulletin*, 18 February 2000.
6. H. T. Hollman, 'The Fatty Acids of Chaulmoogra Oil in the Treatment of Leprosy and Other Diseases', *Archive of Dermatology*, vol. 5, no. 1, pp. 94–101, 1922.

Tu Youyou
1. D. A. Joy, 'Early Origin and Recent Expansion of Plasmodium Falciparum', *Science* (New York, NY), vol. 300, no. 5617, pp. 318–21, 2003.
2. 'Tu Youyou – Biographical', Nobel Media, 2016. [Online]. Available: https://www.nobelprize.org/prizes/medicine/2015/tu/biographical/. [Accessed 26 February 2021].
3. T. Youyou, *From Artemisia annua L. to Artemisinins: The Discovery and Development of Artemisinins and Antimalarial Agents*, Academic Press, 2017.

4. C. Weiyuan, 'Ancient Chinese anti-fever cure becomes panacea for malaria', Bulletin of the World Health Organization, vol. 87, no. 10, pp. 743–4, 2009.
5. T. Youyou, 'The discovery of artemisinin (qinghaosu) and gifts from Chinese medicine', *Nature Medicine*, vol. 17, no. 10, pp. 1217–20, 2011.
6. ibid.
7. T. Youyou, *From Artemisia annua L. to Artemisinins: The Discovery and Development of Artemisinins and Antimalarial Agents*, Academic Press, 2017.
8. T. Youyou, 'The discovery of artemisinin (qinghaosu) and gifts from Chinese medicine', *Nature Medicine*, vol. 17, no. 10, pp. 1217–20, 2011.
9. E. Hsu, 'Reflections on the 'discovery' of the antimalarial qinghao', *British Journal of Clinical Pharmacology*, vol. 61, no. 6, pp. 666–70, 2006.
10. 'Tu Youyou – Biographical', Nobel Media, 2016. [Online]. Available: https://www.nobelprize.org/prizes/medicine/2015/tu/biographical/. [Accessed 26 February 2021].
11. R. Carter and K. N. Mendis, 'Evolutionary and Historical Aspects of the Burden of Malaria', *Clinical Microbiology*, vol. 15, no. 4, pp. 564–94, 2002.

Françoise Barré-Sinoussi
1. 'From discovery to a cure: A conversation with Françoise Barré-Sinoussi', International AIDS Society, [Online]. Available: https://www.iasociety.org/IASONEVOICE/From-discovery-to-a-cure-A-conversation-with-Francoise-Barre-Sinoussi . [Accessed 22 February 2021].
2. 'Françoise Barré-Sinoussi – Biographical', Nobel Media, 2009. [Online]. Available: https://www.nobelprize.org/prizes/medicine/2008/barre-sinoussi/biographical/. [Accessed 21 February 2021].
3. ibid.
4. ibid.
5. Epidemiologic Notes and Reports, Pneumocystis Pneumonia – Los Angeles, 'Morbidity and Mortality Weekly Report', Epidemiologic Notes and Reports, Pneumocystis Pneumonia – Los Angeles, vol. 30, no. 21, 1981.
6. 'A Timeline of HIV and AIDS', HIV.gov, [Online]. Available: https://www.hiv.gov/hiv-basics/overview/history/hiv-and-aids-timeline. [Accessed 22 February 2021].
7. L. K. Altman, 'New homosexual disorder worries health officials', *New York Times*, 11 May 1982.
8. 'Françoise Barré-Sinoussi – Nobel Lecture', Nobel Media, [Online]. Available: https://www.nobelprize.org/prizes/medicine/2008/barre-sinoussi/lecture/. [Accessed 22 February 2021].
9. 'Origin of HIV& AIDS', Avert.org, 30 October 2019. [Online]. Available: https://www.avert.org/professionals/history-hiv-aids/origin. [Accessed 22 February 2021].
10. 'From discovery to a cure: A conversation with Françoise Barré-Sinoussi', International AIDS Society, [Online]. Available: https://www.iasociety.org/IASONEVOICE/From-discovery-to-a-cure-A-conversation-with-Francoise-Barre-Sinoussi. [Accessed 22 February 2021].
11. 'Françoise Barré-Sinoussi – Nobel Lecture', Nobel Media, [Online]. Available: https://www.nobelprize.org/prizes/medicine/2008/barre-sinoussi/lecture/. [Accessed 22 February 2021].

12. F. Barré-Sinoussi and J. C. Chermann, 'Isolation of a T-lymphotropic retrovirus from a patient at risk for acquired immune deficiency syndrome (AIDS)', *Science*, vol. 220, no. 4599, pp. 868–71, 1983.
13. P. Benkimoun and F. Nouchi, 'France, a pioneer against AIDS', *Le Monde*, 13 November 2008. [Online]. Available: https://www.lemonde.fr/planete/article/2008/10/07/prix-nobel-la-france-pionniere-contre-le-sida_1104045_3244.html. [Accessed 21 February 2021].
14. M. Senthilingam, 'HIV discoverer: To develop a cure is almost impossible', CNN Health, 24 July 2015. [Online]. Available: https://edition.cnn.com/2015/07/23/health/francoise-barre-sinoussi-hiv/. [Accessed 21 February 2021].
15. 'Latest global and regional statistics on the status of the AIDS epidemic 2020', UNAIDS, [Online]. Available: https://www.unaids.org/en/resources/fact-sheet. [Accessed 22 February 2021].

Rosalyn Yalow
1. R. J. Anderson, 'Breaking Barriers: The Life and Work of Rosalyn Yalow', *The Pharmacologist*, vol. 59, no. 3, pp. 152–63, 2017.
2. S. Y. Tan and A. Bracha, 'Rosalyn Yalow (1921–2011): Madame Curie from the Bronx', *Singapore Medical Journal*, vol. 60, no. 7, pp. 337–8, 2019.
3. W. Odelberg, 'http://www.nobelprize.org/nobel_prizes/medicine/ laureates/1977/yalow-bio.html', Nobel Media, [Online]. Available: http://www.nobelprize.org/nobel_prizes/medicine/ laureates/1977/yalow-bio.html. [Accessed 27 03 2021].
4. R. J. Anderson, 'Breaking Barriers: The Life and Work of Rosalyn Yalow', *The Pharmacologist*, vol. 59, no. 3, pp. 152–63, 2017.
5. S. Y. Tan and A. Bracha, 'Rosalyn Yalow (1921–2011): Madame Curie from the Bronx', *Singapore Medical Journal*, vol. 60, no. 7, pp. 337–8, 2019.
6. 'Rosalyn Yalow – Biographical', Nobel Media, [Online]. Available: https://www.nobelprize.org/prizes/medicine/1977/yalow/biographical/. [Accessed 06 March 2021].
7. ibid. [Accessed 27 March 2021].
8. ibid. [Accessed 06 March 2021].
9. M. Hargittai, *Women Scientists: Reflections, Challenges, and Breaking Boundaries*, University Oxford Press, 2015.
10. Molecular Endocrinology, 'In Memoriam: Dr. Rosalyn Yalow, PhD., 1921–2011', *Molecular Endocrinology*, vol. 26, no. 5, pp. 713–14, 2012.
11. M. Hargittai, *Women Scientists: Reflections, Challenges, and Breaking Boundaries*, University Oxford Press, 2015.
12. M. Hargittai, 'Mildred S. Dresselhaus', *Candid Science IV*, Imperial College Press, 2004, p. 548.
13. S. Glick, 'Rosalyn Sussman Yalow (1921–2011)', *Nature*, vol. 474, p. 580, 2011.
14. 'In Memoriam: Dr. Rosalyn Yalow, PhD., 1921–2011', *Molecular Endocrinology*, vol. 26, no. 5, pp. 712–13, 2012.
15. C. Holden, 'Yalow Declines Ladies' Award', *Science*, vol. 200, no. 4349, p. 1464, 1978.
16. M. Hargittai, *Women Scientists: Reflections, Challenges, and Breaking Boundaries*, University Oxford Press, 2015.

Chapter 5: Protecting the Earth
Saruhashi Katsuko
1. 'Blazing a Path: Japanese Women's Contributions to Modern Science', The Committee for the Encouragement of Future Scientists, 2001.
2. ibid.
3. Y. Miyake and K. Saruhashi, 'The Carbon Dioxide System in the Ocean', *Papers in Meteorology and Geophysics*, vol. 27, no. 4, pp. 119–28, 1976.
4. K. Saruhashi, 'On the Equilibrium Concentration Ratio of Carbonic Acid Substances Dissolved in Natural Water, A Study on the Metabolism in Natural Waters (II)', *Papers in Meteorology and Geophysics*, vol. 6, no. 1, pp. 38–55, 1955.
5. '5 Things You Should Know About Pioneering Geochemist Katsuko Saruhashi', *TIME* Magazine, 22 March 2018. [Online]. Available: https://time.com/5210207/katsuko-saruhashi-google-doodle/. [Accessed 15 January 2021].
6. 'The Legacy of US Nuclear Testing and Radiation Exposure in the Marshall Islands', US Embassy in the Republic of the Marshall Islands, 15 September 2012. [Online]. Available: https://mh.usembassy.gov/the-legacy-of-u-s-nuclear-testing-and-radiation-exposure-in-the-marshall-islands/. [Accessed 15 January 2021].
7. Y. Miyake, K. Saruhashi and et al, 'Cesium 137 and Strontium 90 in Sea Water', *Journal of Radiation Research*, vol. 2, no. 1, pp. 25–8, 1961.
8. 'Treaty Banning Nuclear Weapon Tests in the Atmosphere, in Outer Space, and Under Water', US Department of State, [Online]. Available: https://2009-2017.state.gov/t/avc/trty/199116.htm. [Accessed 15 January 2021].
9. S. Hatakeyama, 'A Life Story of Saruhashi Katsuko (1920– 2007)', HSSC 528 Gender and Science.

Rachel Carson
1. 'Rachel Carson Biography', US Fish & Wildlife Service, 05 February 2013. [Online]. Available: https://www.fws.gov/refuge/rachel_carson/about/rachelcarson.html. [Accessed 14 February 2021].
2. L. Lear, 'In Memoriam: Mary Scott Skinker', Rachel Carson, [Online]. Available: http://www.rachelcarson.org/mMarySkinker.aspx. [Accessed 14 February 2021].
3. Rachel Carson: Voice of Nature. [Film]. United States: WIFT, 2018.
4. 'Rachel Carson Biography', US Fish & Wildlife Service, 05 February 2013. [Online]. Available: https://www.fws.gov/refuge/rachel_carson/about/rachelcarson.html. [Accessed 14 February 2021].
5. L. Lear, 'In Memoriam: Mary Scott Skinker', RachelCarson.Org, [Online]. Available: http://www.rachelcarson.org/mMarySkinker.aspx. [Accessed 14 February 2021].
6. 'Rachel Carson Biography', US Fish & Wildlife Service, 05 February 2013. [Online]. Available: https://www.fws.gov/refuge/rachel_carson/about/rachelcarson.html. [Accessed 14 February 2021].
7. L. Lear, 'The Next Page: When Rachel Carson set sail', *Pittsburgh Post Gazette*, 21 February 2016. [Online]. Available: https://www.post-gazette.com/opinion/Op-Ed/2016/02/21/The-Nexzt-Page-Rachel-Carson-s-debut/stories/201602210013. [Accessed 14 February 2021].
8. 'Rachel Carson Biography', US Fish & Wildlife Service, 05 February 2013. [Online]. Available: https://www.fws.gov/refuge/rachel_carson/about/rachelcarson.html. [Accessed 14 February 2021].

9. T. Dunlap, *DDT: Scientists, Citizens, and Public Policy*, Princeton University Press, 1981.
10. Rachel Carson: Voice of Nature. [Film]. United States: WITF, 2018.
11. R. Carson, *Silent Spring*, Houghton Mifflin, 1962.
12. ibid.

Wangari Maathai
1. W. Maathai, *Unbowed: My Autobiography*, Arrow Books, 2008.
2. 'Wangari Maathai – Facts', Nobel Media, [Online]. Available: https://www.nobelprize.org/prizes/peace/2004/maathai/facts/. [Accessed 02 March 2021].
3. 'Mau Mau uprising: Bloody history of Kenya conflict', BBC News, 7 April 2011, [Online]. Available: https://www.bbc.co.uk/news/uk-12997138 [Accessed 03 May 2021].
4. Karen Rothmyer, 'The African Airlift', The Nation, 16 September 2009 [Online]. Available: https://www.thenation.com/article/archive/african-airlift/. [Accessed 03 May 2021].
5. W. Maathai, *Green Belt Movement: Sharing the Approach and the Experience*, Lantern Books, 2004.
6. ibid.
7. L. Brown, 'First 'green Nobel' winner, Wangari Maathai dies', *Earth Times*, 26 September 2011. [Online]. Available: http://www.earthtimes.org/politics/first-green-nobel-winner-wangari-maathai-dies/1413/. [Accessed 02 March 2021].
8. ibid.

Chapter 6: Influential Projects and Leadership in STEM
Kate Gleason
1. J. Gleason, *The Life and Letters of Kate Gleason*, RIT Press, 2010.
2. ibid.
3. ibid.
4. N. Bartels, 'The First Lady of Gearing', *Gear Technology*, September/October, pp. 11–17, 1998.
5. H. Bennett, 'Kate Gleason's Adventures in a Man's Job', *The American Magazine*, pp. 42-43, 158-175, Oct 1928.
6. ibid.
7. C. W. Conable, *Women at Cornell: The Myth of Equal Education*, Ithica, NY: Cornell University Press, 1977.
8. History of Graduate Study at Cornell', Cornell University Graduate School, [Online]. Available: https://gradschool.cornell.edu/about/history/. [Accessed 28 March 2020].
9. ibid.
10. *San Fransisco Bulletin*, 19 June 1885.
11. J. Gleason, *The Life and Letters of Kate Gleason*, RIT Press, 2010.
12. J. Gleason, Letter from James, regarding state of business, 1884.
13. W. Gleason, William Gleason to Kate Gleason, 1885.
14. W. Gleason, The Kate Gleason Papers, William Gleason to Kate Gleason, 1885.
15. H. Bennett, 'Kate Gleason's Adventures in a Man's Job', *The American Magazine*, pp. 42–3, 169, October 1928.
16. ibid.

17. ibid.
18. 'Fortnightly Ignorance Club Papers', Rare Books, Special Collections, and Preservation, River Campus Libraries, University of Rochester, 1881-1891. [Online]. Available: https://rbscp.lib.rochester.edu/finding-aids/D201. [Accessed 29 March 2020].
19. J. Gleason, *The Life and Letters of Kate Gleason*, RIT Press, 2010.
20. K. Gleason, Letter from Kate Gleason to James Gleason, 1888.
21. H. Bennett, 'Kate Gleason's Adventures in a Man's Job', *The American Magazine*, pp. 42-43, 171, October 1928.
22. ibid.
23. K. Gleason, 'Letter to the Editor', *New York Times*, 21 May 1910.
24. J. Gleason, *The Life and Letters of Kate Gleason*, RIT Press, 2010.

Caroline Haslett
1. C. Jones, 'Careers and controversy before the First World War', *Nature*, vol. 575, pp. 239–42, 2019.
2. P. Carter, 'The first women at university: remembering 'the London Nine', *Times Higher Education*, 28 January 2018. [Online]. Available: https://www.timeshighereducation.com/blog/first-women-university-remembering-london-nine. [Accessed 21 February 2021].
3. G. R. Rubin, 'Law as a Bargaining Weapon: British Labour and the Restoration of Pre-War Practices Act 1919', *The Historical Journal*, vol. 32, no. 4, pp. 925–45, 1989.
4. 'Report of the War Cabinet Committee on Women in Industry', War Cabinet Committee, 1919.
5. 'Archives Biographies: Dame Caroline Haslett', The Institute of Engineering and Technology, [Online]. Available: https://www.theiet.org/publishing/library-archives/the-iet-archives/biographies/dame-caroline-haslett/. [Accessed 23 January 2021].
6. 'The electrifying life of Caroline Haslett', The Institute of Engineering and Technology, November 2020. [Online]. Available: https://ietarchivesblog.org/2020/12/24/the-electrifying-life-of-caroline-haslett/. [Accessed 23 January 2021].
7. C. Davidson, *A Woman's Work is Never Done*, Chatto & Windus, 1982.

Lillian Gilbreth
1. Kass-Simon G and Farnes P, *Women of Science: Righting the Record*, United States, Indiana University Press, p. 157, 1993.
2. L. Karsten, 'Writing and the Advent of Scientific Management: The Case of Time and Motion Studies', *Scandinavian Journal of Management*, vol. 12, no. 1, pp. 41–55, 1996.
3. C. B. Thompson, 'The Literature of Scientific Management', *The Quarterly Journal of Economics*, vol. 28, no. 3, pp. 506–57, 1914.
4. L. D. Graham, 'Domesticating Efficiency: Lillian Gilbreth's Scientific Management of Homemakers 1924–1930', *Signs*, vol. 24, no. 3, p. 640, 1999.
5. 'Original Films Of Frank B Gilbreth (Part I)', Presented by James S. Perkins in collaboration with Dr Lillian M. Gilbreth and Dr Ralph M. Barnes, [Online]. Available: https://archive.org/details/OriginalFilm. [Accessed 25 October 2020].

6. L. Held, 'Profile of Lillian Gilbreth', Psychology's Feminist Voices Multimedia Internet Archive, 2010. [Online]. Available: http://www.feministvoices.com/lillian-gilbreth/. [Accessed 25 October 2020].
7. P. Edwards and C. Lowndes, 'How this family built life hack culture', VOX, 2 April 2019. [Online]. Available: https://www.youtube.com/watch?v=GeYylII-Nhs. [Accessed 26 October 2020].
8. F. B. Gilbreth and E. G. Carey, *Cheaper by the Dozen*, Thomas Y. Crowell Co., 1948.
9. L. D. Graham, 'Domesticating Efficiency: Lillian Gilbreth's Scientific Management of Homemakers, 1924–1930', *Signs*, vol. 24, no. 3, p. 648, 1999.
10. J. W. Gibson and et al, 'Viewing the Work of Lillian M Gilbreth through the lens of critical biography', *Journal of Management History*, vol. 21, no. 3, pp. 288–308, 2015.
11. L. D. Graham, 'Domesticating Efficiency: Lillian Gilbreth's Scientific Management of Homemakers, 1924–1930', *Signs*, vol. 24, no. 3, p. 661, 1999.

Emily Roebling
1. E. Roebling, 'Essay: A Wife's Disabilities', *Trenton Evening Times*, 1899.
2. D. McCullough, *The Great Bridge*, Simon & Schuster, 1972.
3. A. Padnani and J. Bennett, 'Overlooked', *New York Times*, 8 March 2018. [Online]. Available: https://www.nytimes.com/interactive/2018/obituaries/overlooked.html. [Accessed 11 January 2021].
4. D. McCullough, *The Great Bridge*, Simon & Schuster, 1972.
5. 'New York City Bridge Traffic Volumes', New York City Department of Transportation, New York City, 2016.
6. M. Logan, *The Part Taken by Women in American History*, The Perry-Nalle Publishing Co, 1912.
7. 'About the Brooklyn Bridge', *The New York Sun*, p. 6, 11 June 1891.

Chapter 7: Healthcare for Children
Anna Freud
1. E. Young-Bruehl, *Anna Freud: A Biography*, Yale University Press, 2015.
2. P. Gay, *Freud: A Life for Our Time*, London: J. M. Dent & Sons, 1988.
3. P. Gay, *Reading Freud*, London: Yale University Press, 1990.
4. E. Young-Bruehl, *Anna Freud: A Biography*, Yale University Press, 2008.
5. ibid.
6. N. Midgely, 'The Matchbox School' (1927–1932): Anna Freud and the Idea of a 'Psychoanalytically Informed Education', *Journal of Child Psychotherapy*, vol. 34, no. 1, pp. 23–42, 2008.
7. 'Anna Freud', The School of Life, [Online]. Available: https://www.theschooloflife.com/thebookoflife/the-great-psychoanalysts-anna-freud/. [Accessed 28 March 2021].
8. '1941–1990; A Breakthrough in Practice', The Anna Freud Centre, [Online]. Available: https://www.annafreud.org/about-us/our-history/. [Accessed 28 March 2021].
9. J. Goldstein, A. Freud and A. J. Solnit, *Beyond the Best Interests of the Child*, Free Press, 1973.

10. '1991 – Present day, Her legacy', The Freud Centre, [Online]. Available: https://www.annafreud.org/about-us/our-history/. [Accessed 28 March 2021].

Mammie Phipps Clark
1. US Reports: Brown v. Board of Education, 347 US 483 (1954), 1953.
2. US Reports: Brown v. Board of Education, 349 US 294 (1955), 1954.
3. W. Warren, *Black Woman Scientists in the United States*, Indiana University Press, 1999.
4. 'The Significance Of 'The Doll Test', NAACPLDF, [Online]. Available: https://naacpldf.org/ldf-celebrates-60th-anniversary-brown-v-board-education/significance-doll-test/. [Accessed 03 October 2020].
5. R. E. Horowitz, 'Racial Aspects of Self-Identification in Nursery School Children', *The Journal of Psychology Interdisciplinary and Applied*, vol. 7, pp. 91–9, 1938.
6. K. Clark and M. Clark, 'The development of consciousness of self and the emergence of racial identification in Negro pre-school children', *The Journal of Social Psychology*, vol. 10, pp. 591–599, 1939.
7. A. C. Gallattee, 'The Negro psyche: fact, fiction and fantasy', *Journal of the National Medical Association*, vol. 61, no. 2, pp. 119–29, 1969.
8. W. Warren, *Black Woman Scientists in the United States*, Indiana University Press, 1999.

Virginia Apgar
1. 'Bubbles', Gapminder, [Online]. Available: https://www.gapminder.org/tools. [Accessed 16 February 2021].
2. H. Rosling, *Factfulness*, Sceptre, 2018.
3. 'Changing the Face of Medicine, Dr Virginia Apgar', National Library of Medicine, [Online]. Available: https://cfmedicine.nlm.nih.gov/physicians/biography_12.html. [Accessed 16 February 2021].
4. M. Finster and M. Wood, 'The Apgar score has survived the test of time', Anesthesiology, vol. 102, no. 4, pp. 855–7, 2005.
5. 'Virginia Apgar: Biographical Overview', National Library of Medicine, [Online]. Available: https://profiles.nlm.nih.gov/spotlight/cp/feature/biographical-overview. [Accessed 18 February 2021].
6. J. M. Doyle, S. Echevarria and W. P. Frisbie, 'Race/ethnicity, Apgar and infant mortality', *Population Research and Policy Review*, vol. 22, pp. 41–64, 2003.
7. V. Apgar, 'A Proposal for a New Method of Evaluation, of the Newborn Infant, Current Researches', *Anesthesia and Analgesia*, vol. 32, p. 260, 1953.
8. ibid p. 261.
9. ibid p. 262.
10. L. S. James, 'Fond memories of Virginia Apgar', *Pediatrics*, vol. 55, no. 1, pp. 1–4, 1975.

Helen Taussig
1. J. Van Robays, 'Helen B. Taussig (1898-1986).', *Facts, Views & Vision in ObGyn*, vol. 8, no. 3, pp. 183–7, 2016.
2. G. L. Goodman, A gentle heart: the life of Helen Taussig, Yale Medicine Thesis Digital Library, 1983.
3. J. Van Robays, 'Helen B. Taussig (1898-1986)', *Facts, views & vision in ObGyn*, vol. 8, no. 3, pp. 183–7, 2016.

4. 'Dr Helen Brooke Taussig', National Library of Medicine, [Online]. Available: https://cfmedicine.nlm.nih.gov/physicians/biography_316.html. [Accessed 28 March 2021].
5. J. Harvey and M. Ogilvie, *The Biographical Dictionary of Women in Science*, Routledge, 2000.

Chapter 8: Understanding Our World
Inge Lehmann
1. S. Bush, 'Discovery of the Earth's Core', *American Journal of Physics*, vol. 48, pp. 705–724, 1980.
2. E. Hjortenberg, 'Inge Lehmann's work materials and seismological epistolary archive', *Annals of Geophysics*, vol. 52, no. 6, pp. 679–98, 2009.
3. M. Kölbl-Ebert, 'Inge Lehmann's paper: 'P'', *Episodes*, vol. 24, no. 4, 2001.
4. Unknown, 'Inge Lehmann', Unknown, [Online]. Available: https://reference.jrank.org/biography-2/Lehmann_Inge.html. [Accessed 1 March 2020].
5. B. A. Bolt, 'Inge Lehmann. 13 May 1888—21 February 1993', Biographical Memoirs of Fellows of the Royal Society, vol. 43, no. 287, 1997.
6. 'Inge Lehmann: Discoverer of the Earth's Inner Core', American Museum of Natural History, [Online]. Available: https://www.amnh.org/learn-teach/curriculum-collections/earth-inside-and-out/inge-lehmann-discoverer-of-the-earth-s-inner-core. [Accessed 1 March 2020].
7. Unknown, 'Inge Lehmann', Unknown, [Online]. Available: https://reference.jrank.org/biography-2/Lehmann_Inge.html. [Accessed 01 March 2020].
8. I. Lehmann, 'The earthquake of 22 III 1928, Gerlands Beitr', Geophysics, vol. 28, no. 151, 1930.
9. 'Inge Lehmann: Fellow Details', The Royal Society, [Online]. Available: https://doi.org/10.1098/rsbm.1997.0016 [Accessed 01 March 2020].
10. B. Bolt, 'Inge Lehmann', *Physics Today*, p. 61, 1994.
11. M. Kölbl-Ebert, 'Inge Lehmann's paper: 'P'', *Episodes*, vol. 24, no. 4, 2001.
12. Bruce Bolt, 'Inge Lehmann', *Physics Today*, p. 61, 1994.
13. B. S. Lady Jeffreys, 'Inge Lehmann: Reminiscences', *Quarterly Journal of the Royal Astronomical Society*, vol. 35, no. 2, p. 231, 1994.

Maria Goeppert Mayer
1. 'Maria Goeppert Mayer – Biographical', Nobel Media, [Online]. Available: https://www.nobelprize.org/prizes/physics/1963/mayer/biographical/. [Accessed 28 February 2021].
2. M. Hargittai, *Women Scientists: Reflections, Challenges, and Breaking Boundaries*, Oxford University Press, 2015.
3. K. Johnson, 'Science at the Breakfast Table', *Physics in Perspectives*, vol. 1, pp. 22–34, 1999.
4. R. Sachs, 'Maria Goeppert Mayer 1906–1972: A Biographical Memoir', Biographical Memoirs. *National Academy of Sciences*, 1979.
5. 'Maria Goeppert Mayer: Revisiting Science at Sarah Lawrence College', Sarah Lawrence College, [Online]. Available: https://www.sarahlawrence.edu/archives/exhibits/maria-goeppert-mayer-exhibit/. [Accessed 04 March 2021].
6. ibid.

7. 'Maria Goeppert Mayer – Women Who Changed Science', Nobel Media, [Online]. Available: https://www.nobelprize.org/womenwhochangedscience/stories/maria-goeppert-mayer. [Accessed 06 March 2021].
8. M. Hargittai, *Women Scientists: Reflections, Challenges, and Breaking Boundaries*, Oxford University Press, 2015.
9. 'Maria Goeppert Mayer – Women Who Changed Science', Nobel Media, [Online]. Available: https://www.nobelprize.org/womenwhochangedscience/stories/maria-goeppert-mayer. [Accessed 06 March 2021].

Marie and Irène Curie
1. 'Marie Curie – Polish Girlhood (1867–1891) Part 1', American Institute of Physics, [Online]. Available: https://history.aip.org/exhibits/curie/. [Accessed 30 June 2020].
2. W.A. Wierzewski, 'Mazowieckie korzenie Marii' [Maria's Mazowsze Roots]', *Gwiazda Polarna*, vol. 100, no. 13, pp. 16–17, 2008.
3. T. Estreicher, 'Curie, Maria ze Skłodowskich', *Polski słownik biograficzny*, vol. 4, p. 111, 1938.
4. M. Hargittai, 'Husband and Wife Teams', *Women Scientists: Reflections, Challenges and Breaking Boundaries*, Oxford University Press, 2015, p. 10.
5. 'Marie Curie the scientist', Marie Curie Cancer Care, [Online]. Available: https://www.mariecurie.org.uk/who/our-history/marie-curie-the-scientist. [Accessed 30 June 2020].
6. A. R. Coppes-Zantinga and M. J. Coppes, 'Marie Curie's contributions to radiology during World War I', *Medical and Pediatric Oncology*, vol. 31, no. 6, pp. 541–3, 1998.
7. Z. Tsjeng, 'Medicine and Psychology, Irène Joliot-Curie', *Forgotten Women: The Scientists*, Octopus Publishing, 2018.
8. A. Niroomand-Rad, 'Oral Histories: Hélène Langevin-Joliot', American Institute of Physics, 13 April 2003. [Online]. Available: https://www.aip.org/history-programs/niels-bohr-library/oral-histories/31377. [Accessed 06 July 20].

Gerty Cori
1. S. B. McGrayne, *Nobel Prize Women in Science: Their Lives, Struggles and Momentous Discoveries*, 2nd edition, NJ: Carol Publishing Group, 1993.
2. M. Hargittai, 'Husband and Wife Teams', *Women Scientists: Reflections, Challenges and Breaking Boundaries*, Oxford University Press, 2015, p. 14.
3. 'Gerty Cori – Facts', Nobel Media, [Online]. Available: https://www.nobelprize.org/prizes/medicine/1947/cori-gt/facts/. [Accessed 23 June 2020].
4. M. Hargittai, 'Husband and Wife Teams', *Women Scientists: Reflections, Challenges and Breaking Boundaries*, Oxford University Press, 2015, p. 16.

Annie Easley
1. S. Johnson, 'NASA Headquarters Oral History Project: Annie J. Easley', NASA, 21 August 2001. [Online]. Available: https://historycollection.jsc.nasa.gov/JSCHistoryPortal/history/oral_histories/NASA_HQ/Herstory/EasleyAJ/EasleyAJ_8-21-01.htm. [Accessed 19 January 2021].
2. ibid.
3. ibid.

4. A. K. Mills, 'Annie Easley, Computer Scientist', NASA, 7 August 2017. [Online]. Available: https://www.nasa.gov/feature/annie-easley-computer-scientist. [Accessed 19 01 2021].

Constance Tipper, Marie Gayler and Marion McQuillan
1. A. Sachidhanandham, 'Interesting dimensions of Ahimsa silk (Vegan silk)', *Indian Textile Journal*, pp. 108–12, 2020.
2. H. C. H. Carpenter and C. F. Elam, 'Crystal Growth and Recrystallisation in Metals', *Nature*, vol. 106, pp. 312–15, 1920.
3. G. I. Taylor, 'The Mechanism of Plastic Deformation of Crystals. Part I. Theoretical', Proceedings of the Royal Society of London. Series A, Containing Papers of a Mathematical and Physical Character, vol. 145, no. 855, pp. 362–87, 1934.
4. K. Zappas, 'Constance Tipper Cracks the Case of the Liberty Ships', *Journal of Materials*, vol. 67, no. 12, pp. 2774–6, 2015.
5. ibid.
6. N. C. Baker, '25: Marie Laura Violet Gayler' Magnificent Women, 25 March 2019. [Online]. Available: https://www.magnificentwomen.co.uk/engineer-of-the-week/25-marie-laura-violet-gayler. [Accessed 13 March 2021].
7. A. J. Murphy, 'Obituary: Marie Laura Violet Gayler', *Nature*, vol. 263, pp. 535–6, 1976.
8. E. H. Greener, 'Amalgam: Yesterday, Today, and Tomorrow', *Operative Dentistry*, vol. 4, pp. 24–35, 1979.
9. R. Greg, *Titanium* (First ed.), New York, NY: The Rosen Publishing Group, 2008.
10. 'Material property charts', ANSYS/Granta, [Online]. Available: https://www.grantadesign.com/education/students/charts/. [Accessed 27 March 2021].

Index

5S 99

Acquired immune deficiency syndrome 68
African American Students Foundation 81
Algebra 30, 86
Algorithm 9–10, 30
American Birth Control League 32
American Institute for the Prevention of Blindness 48
American Medical Association 33, 50
American Railway Engineering Association 29
Amino acids 55, 59
Anaesthesiology 108
Analytical engine 7
Anschluss 12, 105
Antibody 71
Antigen 68, 71
Antimalarial 63, 65–6
Apgar Scoring System 108
Artificial radioactivity 122

Babbage, Charles 7–8
Bacteriology 27
Baltimore and Ohio Railroad 28
Becquerel, Henry 121
Berson, Solomon 70–1
Biodiversity 78
Birth control 31–5
Birth Control Clinical Research Bureau 32
Birth defects 109
Blalock – Thomas – Taussig Shunt 110
Bletchley Park 1–3
Bohr, Niels 12

Boltzmann, Ludwig 11
Bombe 1–2
British Coal Utilisation Research Association 4
Brooklyn Bridge 100, 102

Cancer 45, 58–60, 66–7, 70, 75, 79, 122, 124
Carbohydrate 123–4
Carbon dioxide (CO_2) 73–4
Catalysis 25
Cataract 45, 48–9
Centaur 125–6
Central heating 19
Chaulmoogra oil 61–2
Chemotherapy 59
Childbirth 31
Chimera 37
China National Committee of Science and Technology 65
Churchill, Winston 2
Coal 4
Crack (cocaine) 43
Code (computing) 9
Code (cypher) 1–2
College of Physicians and Surgeons, USA 108
Common business-oriented language 9
Computers (job role) 125–6
Computers (machine) 7–9, 74, 115, 125
Comstock Act 32
Conservation 38, 75, 77–8
Consumer Guidance Society of India 58
Contraceptives 33
Cori Cycle 124
Crick, Francis 3, 5
Cryptanalysis 1

Cultural Revolution 64–5
Cypher 1–2

DDT 78
Defence mechanism 104
Deforestation 80
Difference engine 7–8
Diphtheria 27
DNA 3–5, 37–8, 59, 66–7, 125
Du Bois, W.E.B 32

Earthquake 113–4
Ecology 77
Ego 104
Electrical Association for Women 93
Electricity 14, 41, 90, 93
Electron 117
Energy level 118–19
Engineering 8, 17, 28–9, 86–7, 89, 90–93, 94–5, 97, 99, 100, 102, 128–32
Enigma code 1–2
Enzyme 67–8, 71, 124–5
Epidemic 43, 49, 67–8
Ergonomics 94, 96
Eugenics 33
Experimental physics 40–1

Family planning 32–3
Fermi, Enrico 12, 116–7, 119
Ford, Henry 22, 28
Fossil fuels 10

Gage, Matilda Joslyn 17
Galton, Francis 33
Gandhi, Mahatma 54–5, 57
Genetics 36–8
Geodetical Institute of Denmark 114
Glucose 123–4
Glycogen 124
Gosling, Raymond 4–5
Green Belt Movement 82

Habitat 10
Harvard Mark I computer 9
Heroin 43, 68
HIV 68–9

Hopper, Grace 9
House of Wisdom, Baghdad 30

Immune response 68, 71
In vitro fertilisation 36
Incarceration 27
Industrial engineering 94–5
Institute of Animal Genetics 37
Institute of Chinese Materia Medica 64
Institute of electrical engineers 93
Institut Pasteur 66
Insulin 70–1, 125
Isotope 70, 117–18

Japanese Christian Students Association 27

Kaposi's sarcoma 67
Kevlar(r) 25
King, Martin Luther, Jr 33, 46, 48, 59

Lamarr, Hedy 8
Legumes 57
Lehmann discontinuity 115
Leprosy 60–2
Leukaemia 27, 59–60, 67–8, 123

Magic numbers 118–9
Malaria 51, 63–6
Management 87, 95–9
Manufacturing 20–1, 25, 84, 99, 131
March of Dimes Prize 39
Marine biology 75–6
Matchbox School 104
Medical physics 69
Medical Research Council 4, 37
Mental health 105–106, 120
Meteorological Research Institute, Japan 73
Methotrexate 59
Microstructure 4, 25
Montagnier, Luc 67–8
Musa al-Khwārizmī, Muhammad 30

NASA 125–8
National Association for the Advancement of Colored People 32, 42

National Council of Women in
  Kenya 82
National Medical Association 42
Neera 57–8
Neonatal 109
Nerve growth factor 125
Neutron 12, 117–18
Newton, Isaac 29, 39–41
Newtonian physics 39–40
Nightingale, Florence 15
Nissei 25, 27
Nobel Laureate 54, 56, 66, 72, 116
Nobel Prize 3, 6, 14, 56, 60, 63, 66,
  68–9, 72, 119–25
Nuclear fission 10, 14
Nucleus (atomic) 10, 13, 115, 117–19
Nucleus (cell) 67
Nutrients 56

Obstetrics 52, 108
Oncology 58–60, 122
Ophthalmology 45–9
Otto Hahn 11, 14

Partition of India 57
Periodic table 12, 117
Pesticide 78
Pharmaceutical science 63
Photo 51 5–6
Planck, Max 11
Planned Parenthood 35
Pollen 27
pollution 74, 81
Polonium 121–2
Polymer 24
Poor People's Campaign 46
Pregnancy 31, 33
Project 523 64–5
Protein 55–6, 59, 71
Psychiatry 41, 43, 94
Psychoanalysis 103, 106
Pulses 55

Quantum physics 116

Radiation 11, 74, 117, 121–3
Radioimmunoassay (RIA) 71

Radioisotope 71–2
Radium 13, 121–2
Raman, Chandrasekhara Venkata 54–5
Retrovirus 66–8
RNA 59, 66
Roswell Cancer Institute 124
Royal Institute of Science, India 57
Rufaida Al-Aslamia Prize 15

Saruhashi Prize 75
Satyagraha 55
Seacole, Mary 15
Search engine 9
Segregation 106, 108
Seismograph 113–14
Seismology 112–14
Sex education 31
Society of Industrial Engineers 98
Somerville, Mary 7, 79
Soy 53
Spärck Jones, Karen 9
Stem cells 38
Strassmann, Fritz 12–14
Student National Medical
  Association 42
Sustainable 10, 18, 78, 131

Taylor, Frederick 95–7
Telomere 125
Test Ban Treaty 74
Tetralogy of Fallot 110
Thalidomide 111
The doll study 106
The Great Depression 76, 99
The Manhattan Project 14, 117
The Negro Project 34
The New York Call 31
The Royal Society 36, 129
The Woman Rebel 32
Time and motion studies 95
Time curve 114
Tobacco mosaic disease 6
Traditional Chinese medicine
  (TCM) 63–4
Transuranium elements 12, 117
Tuberculosis 26, 63, 110
Turing, Alan 1, 3

UK Government Code and Cypher
   School 1
US Department of Agriculture 53
US Department of Defence 8
US Fish and Wildlife Service 77
US Marine Laboratory 76

Vaccination 26
Virology 66
Virus 6, 36, 66–8, 71
Vitamin B12 125
Vitamin 56–8, 71

Warnock Committee 38
Watson, James 3, 5
Wilkins, Maurice 3–6
Women's Engineering Society
   (WES) 92–3
Work triangle 97
World Health Organisation 59

X-ray 4–5, 122, 125

Zoology 36, 42, 76, 81, 108–109